ENGINEERING HYDROLOGY

Jaromír Němec

McGRAW-HILL • LONDON

NEW YORK • SYDNEY • TORONTO
MEXICO • JOHANNESBURG

Published by:

McGRAW-HILL PUBLISHING COMPANY LIMITED

MAIDENHEAD ● BERKSHIRE ● ENGLAND

07 094153   X

Authorised translation from the first Czech language edition
of Inženýrská hydrologie first published by
SNTL, Prague

ⓒ 1964 doc. ing. Jaromír Němec, CSc.

PRINTED AND BOUND IN CZECHOSLOVAKIA

ENGINEERING
HYDROLOGY

Consulting Editor
Professor P. B. MORICE, University of Southampton

CONTENTS

PREFACE

This book has been written for engineers in all branches of engineering and, in particular, for civil and agricultural engineers concerned with the planning, design and construction of water resources development projects.

Complicated problems of hydrological analysis are generally entrusted to specialized institutions which, in many countries — and in particular in Czechoslovakia, where this book was written — are the hydrological services in charge of the collection, transmission and processing of basic hydrological data. If such a service exists, it saves the engineer the labour of collecting and analysing his own data; but the responsibility for using the hydrological design data correctly in his designs will always rest with the engineer and he must therefore have sufficient knowledge of the methods by which data are collected and analysed. In addition, the simpler hydrological computations will often be performed by the engineer himself. Therefore, while describing the methods for collection and analysis of hydrological data which should be generally known, this book concentrates on details of simple computing procedures which are accessible to all engineers in consulting offices and in the field.

In revising and editing the English translation, the author has deliberately conserved an approach based on often little-known central and eastern European practices. At the same time, he has kept in mind that the book must be valid and useful for other countries too and so has included descriptions of methods widely used elsewhere in the world. Current practices in the United Kingdom are also described and, to point up the book's international character, technical publications of international organizations active in this field have been widely drawn upon.

The greater part of the book is concerned with atmospheric and surface water. While the part of the hydrological cycle concerned with sub-surface water – soil and groundwater – is taken into full account when it influences the processes of surface water hydrology, the specific problems of groundwater are not covered. In view of the particular character and importance of this subject, the author feels that it warrants separate and exhaustive treatment which is beyond the scope of this book. The references therefore direct the reader to specialized literature on hydrogeology and hydraulics of groundwater.

This book offers the engineer practical guidance on the computation of atmospheric and surface water hydrological design data for engineering projects. When preparing such a project, a comprehensive report is usually made of the hydrological characteristics of a stream and its basin consisting of the following items (for ease of reference the parts of the book in which information may be found for the preparation of such a report are indicated):

(a) Physical and geographical characteristics of the basin (see chapter 3, pages 125 – 126, chapter 4, page 169).

(b) Climatic characteristics of the basin (chapter 3, pages 58 – 86, chapter 4, pages 175 – 211).

(c) Hydrographic description of the basin (chapter 3, pages 125 – 126, chapter 4, pages 169 – 175).

(d) Information on records of hydrological and climatic data in basins with an evaluation of their reliability (chapter 3, pages 48 – 51).

(e) Character of the main stream, its sources, stream-bed variations in time and space, stage-discharge relations (chapter 4, pages 211 – 220).

(f) Long-term average annual flow and annual flow with a probability of per cent (chapter 4, pages 222 – 228).

(g) Variations of runoff during a year, flow duration curves of daily flow (chapter 4, pages 228 – 234).

(h) Extremes of flow: flood flow (chapter 4, page 235 – 260), minimum flow (chapter 4, pages 261 – 263).

(i) Winter regime, icing (if appropriate) (chapter 1, pages 38 – 40).

(j) Suggestions for further studies or additional surveys.

(k) Summary of data used and depositories of data records.

1   Hydrological Data

1.1   Data requirements for projects

It is not hydrologists who need hydrological data primarily, but engineers responsible for the construction of water resources projects, whether related to hydroelectric power, farming, or industrial and municipal water supply, or in the road, railroad and building industries.

In Czechoslovakia and in several other countries of central and eastern Europe, the data required for water resources development projects are prescribed by standard regulations appropriate to different fields. The regulation for reservoir and dam projects, for example, requires that the following hydrological data be available: basin area, average annual areal precipitation on the basin, average annual flow (yield), average annual duration curve of daily mean discharges, flood peak discharges of recurrence intervals from 1 to 100 or more years. Additional data are required for the evaluation of the river flow regulation by reservoirs; average monthly discharges for typical years for the reservoir storage function, statistical parameters, such as coefficient of variation and skewness, evaporation data for free water evaporation in the basin, data on suspended sediment and bed-load transport, and icing conditions in the basin (freezing, thawing).

Standards of the Czechoslovak Ministry of Agriculture require the following data to be used in irrigation, drainage and small river-development projects: basin area and its physiographical description; average annual precipitation over the basin and its monthly distribution; a series of annual precipitations from the nearest hydrometeorological station; average annual snow pack depth and its water equivalent before the period of thaw; maximum daily storm rainfall; series of storm rainfall intensities of a duration from 10 min to 24 h of frequencies from 1 to 100 years; average annual temperature; average monthly temperature; average monthly saturation deficit of air moisture; data on pan and lake evaporation (monthly maxima); average annual flow (yield); annual flows of different frequency; average monthly flows; duration curve of daily mean discharges; flood peak flows of different frequencies; and minimum discharges.

1.2   Data for hydrometeorological studies

The internationally recognized Guide to Hydrometeorological Practices, published by the *World Meteorological Organization* (1965) in Geneva, indicates the following data which are most frequently needed for hydrometeorological studies:

amount of precipitation—annual, monthly, daily storm;

data on precipitation intensity and frequency—for frequencies from 2 to 50 years and durations from 5 min to 72 h;

variability of precipitation from year to year;

frequency and duration of droughts;

runoff—annual and monthly volumes, flood peaks;

evapotranspiration—actual annual (precipitation minus runoff);

evaporation from water surface—annual and monthly;

variability of annual free-water evaporation;

vapour pressure or dew point—mean annual and monthly;

temperatures of water surface—mean monthly;

precipitable water in atmosphere—mean annual and monthly;

short-wave radiation—total incoming on a horizontal surface—mean annual and monthly;

net radiation balance at earth surface;

air temperature in screen;

soil moisture and soil moisture deficit.

1.3   Data on groundwater

Data on groundwater may serve the following purposes:
(a) Evaluation of the resource for planning purposes.
(b) Study of interrelation of surface water and groundwater as an essential component of hydrological balance calculations of a basin.
(c) Basis for projects in groundwater exploitation for water supply (domestic, municipal, industrial, agricultural), soil stability, mining, waste disposal, land subsidence, etc.
These data have two principal characteristics:
(a) Variability in quantity, the total amount of groundwater yielded, and the rate of flow of groundwater towards points of discharge.
(b) Variability in the dissolved mineral content. While suspended sediment, organic content, and radioactivity are either minor problems in the sub-surface environment or, as in the case of water temperature, relatively invariable in some countries, they may acquire considerable importance in arid and semi-arid zones, particularly with regard to irrigation schemes.
The problem of groundwater is, however, complex, highly specialized, and treated mostly by hydrogeologists. This book will therefore be concerned with it only if it directly influences the surface water.

1.4   Categories of reliability of data

Much of the above data may be found in published records which have been computed directly from field measurements and checked for quality but, for other data, hydrological analysis and computations are necessary. The reliability of data can also be classified. The following classes of reliability are used in Czechoslovak standards:
Class I: data from direct observations and measurements directly on or near the site of the project, on the river concerned, by one or more hydrometric stations.
Class II: data extrapolated from class I data by analogy, for basins analogous by area, geological, geomorphological and climatic conditions.
Class III: data extrapolated as above but with no analogy in one or more of the conditions mentioned above.

Class IV: data computed by empirical formulae.

The above reliability classes are particularly applied to average annual flow, average duration curves of daily flows and flood peak flows of different frequencies.

Data of a certain class of reliability may be used only in projects of corresponding importance, evaluated with respect to capital investment in them. Thus, for water resources projects connected with agriculture,

Table 1

Project	Category			
	a	*b*	*c*	*d*
Drainage	up to 0.10 km²	up to 2.00 km²	up to 20.00 km²	over 20.00 km²
Irrigation	up to 0.05 km²	up to 1.00 km²	up to 20.00 km²	over 20.00 km²
Irrigation with fertilizers	—	up to 1.00 km²	up to 20.00 km²	over 20.00 km²
Water supply and drainage canals	up to 1 km	up to 10 km	over 10 km	
River regulation	basin up to 10 km²	basin up to 50 km²	basin up to 100 km²	basin* up to 150 km²
Reservoirs	—	up to 0.10 km² of water surface	over 0.10 km² of water surface	—
Farm ponds	all			
Roads	serving up to 0.10 km²	up to 2.00 km²	up to 20.00 km²	over 20.00 km²
Conservation watershed management	—	simple	complex	

* As regards basins over 150 km², river regulations and reservoirs above a certain limit are not specified in standards of the Ministry of Agriculture, since they are in the jurisdiction of the Water Resources Authorities.

the classification of projects shown in Table 1 has been established in Czechoslovakia. While there is no direct instruction as to which class of data mentioned above applies to each category, categories *a* and *b* will use mostly data of class II and III, while categories *c* and *d* will require data of class I and II (*Table 1*).

The Czechoslovak standards require that all hydrological data for basins above 5 km^2 be supplied or corroborated by the *Hydrological Service* (Hydrological Departments of the National Hydrometeorological Institute).

A similar practice exists in the USSR and several other countries of eastern Europe. Nevertheless, the practising engineer is and will always be faced with the problem of evaluation of the data, even if these are supplied by the hydrologic or meteorologic service.

2 Basic Hydrological Concepts

2.1 The basin, the hydrological cycle, and its balance

(a) Hydrology is the science of the occurrence and cycle of water on earth. Because of the many forms and ways in which water occurs, hydrology may be divided into several different branches. There are many ways of making such divisions. A classical division assumes that the branch which deals with water in the atmosphere may be called *hydro-meteorology;* with water as it flows over the earth's surface—*potamology;* with water in natural or artificial reservoirs—*limnology;* with ground-water—*geohydrology;* with soil moisture—*pedohydrology.*

Hydrography is concerned with regular observations and descriptions of a hydrological network (rivers, streams, ponds, lakes, etc.), while *hydrometry* deals with the measuring of the water level, discharges and other quantitative aspects of hydrological elements.

Some of the more recent divisions consider hydrometeorology as the part of hydrology concerned with atmospheric and surface water, including only those aspects of sub-surface water closely related to surface water processes (Bruce, Clark, 1968). Sub-surface water hydrology is also often divided into unsaturated and saturated zones. The latter may be called hydrogeology, providing that it is primarily concerned with the geological environment in which the hydrological processes occur.

Many parts of hydrology are closely interrelated with other sciences, such as meteorology, hydraulics, physical geography, etc., and as a result many boundary, interdisciplinary fields exist.

Another approach to classifications of hydrology recognizes *scientific hydrology* as concerned with purely academic aspects, while *engineering hydrology* is to devote its attention to practical application and data computation for engineering purposes.

It is obvious that no one classification can reflect the manifold facets of the water cycle on the earth. If confused, the reader is directed to the definitions of the *International Glossary of Hydrology* (WMO, 1969), since only definitions of the terms used in such classifications can express their boundaries and meanings.

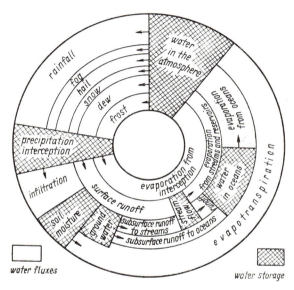

Fig. 2.1. Hydrological cycle

Water on earth and in the atmosphere is known as the *hydrosphere*. The hydrosphere obviously includes the water in the oceans, which is studied by the science of *oceanography*. This book is not intended to consider the subject of water in the oceans. Thus, the part of hydrology treated here would be more precisely defined as hydrology of the continents or the land phase of the hydrological cycle. Furthermore, as already indicated, the emphasis will be concentrated on surface water.

Water constantly evaporates from the surface of bodies of water, from plants and from the soil. It rises into the atmosphere where it condenses in the form of rain which falls on the earth's surface where it is in turn absorbed or flows over or under it and once more evaporates. This constant motion is known as the hydrological cycle *(see Fig. 2.1)*; its source of energy is the radiation of the sun.

(b) The elementary area in which the hydrological cycle can be expressed numerically is called the *basin*. It is a region which is closed, hydrologically speaking, which means that there is no surface or underground flow of water into it and all precipitation excess which falls on its surface flows off through its principal stream or evaporates. A catchment area, the boundaries of which are set by a line called a *watershed divide*, always corresponds to a cross-section of this stream. The catchment for sub-surface water sometimes differs from the one for surface water but, in most cases, in view of the minor difference, we also consider the orographic (terrain) divide as being the groundwater watershed divide.

The hydrological cycle for a basin may be expressed quantitatively by the so-called *hydrological balance* or *balance equation;*

$$P = R + E \pm S \qquad (2.1)$$

in which: P represents the amount of water [m³] which falls as precipitation during a given period on the surface of the basin,

$R =$ amount of water [m³] which flows off the surface (through the closing profile of the basin) or under the surface (runoff)

$E =$ amount of water [m³] which evaporates from water surfaces, from soil and plants in the basin (climatic or total evaporation),

$S =$ amount of water [m³] which has increased or decreased the storage at the surface (ponds, lakes) or of sub-surface water in the basin.

It is very important for all factors of the equation to be taken during the same period of time. The basic hydrological balance period is the *hydrological or water year*. In Czechoslovakia, it begins on 1 November of the previous year and ends on 31 October of the current calendar year. In the United States it runs from 1 October to 30 September, in the United Kingdom from 1 September to 31 August. It varies in other countries with the different climatic conditions in the various parts of the world.

It is set in this way so that all precipitation which has fallen within this period (including ice and snow) will also run off.

The *balance equation* and calculations which are to determine some or all of its factors at different periods of time and for different frequencies of occurrence are the substance of hydrological analysis.

2.2 Humidity, evaporation, and precipitation

Water vapour in the air is called its humidity. Air can contain, at a given temperature and *barometric pressure*, only a certain quantity of water vapour. *Air temperature* is the result of the warming of air from the earth's surface (land and ocean surfaces) which is itself warmed by the sun.

The ratio of the amount of sun's radiation falling on the earth's surface to the amount of radiation reflected by the surface is known as *albedo* (Ad). For soil it amounts to 0.08 to 0.30, for forests 0.05 to 0.18, for water 0.05 to 0.55 and for snow 0.78.

Air may change in temperature because of its movement. If air rises high, it expands and cools. If it is at the same time saturated with water vapour, the vapour condenses, although the air is thereby partially warmed. The opposite process occurs when the air descends.

In Czechoslovakia temperature is generally measured three times a day or else recorded continuously by automatic recording instruments (see section 3). From these thrice-daily observations, the mean daily (t °C [d]), mean monthly (t °C [m]), mean annual (for a given year) (t °C [y]) temperatures, and the average annual temperature for more years (Mt °C [1920 − 1960]) which characterizes a climate, are calculated by arithmetical averages. In some countries, the mean daily temperature is not computed from measurements at given times, but from the daily maximum and minimum, measured by a maximum/minimum thermometer. The differences resulting from the use of one or the other method may sometimes be significant (see section 3.6c).

If relevant temperatures measured at meteorological stations are marked on maps and points with the same temperatures (during the same period) are joined by lines, a map of *isotherms* results. Air pressure changes with elevation above sea level and with the general circulation of the atmosphere, which causes pressure highs and lows. The average atmospheric pressure at sea level is, in SI units, 1 013 250 dynes/cm^2, or 1013 millibars (mb) which, in traditional barometric pressure units, corresponds to 760 mm of mercury (mm Hg).

The warmer the air, the more water vapour it can absorb. If the air becomes saturated at a given temperature and pressure, it reaches maximum moisture E_m [mm Hg or mb] or A [g/m^3]. Since the air, including water vapour, is a mixture of gases and, according to laws of physics, each of them has its own pressure independent of pressures of the others, we can thus express air moisture as absolute humidity E_m in units representing the pressure of water vapour [mm Hg, or mb]. It can also be given as A in grammes of water per cubic metre of air [g/m^3]. Relative humidity of air is the ratio of actual humidity e to maximum humidity E at that temperature and pressure:

$$r = \frac{e}{E_m} \cdot 100 \qquad [\%] \qquad\qquad (2.2)$$

The difference between actual and maximum humidity is the *saturation deficit* d

$$d = E_m - e \qquad [\text{mm Hg, or mb}] \qquad\qquad (2.3)$$

The temperature at which saturation is reached under a given air pressure is known as the *dew point* [t°]. In Appendix 6, the so-called *psychrometric chart* is given which indicates this pressure as a function of the temperature.

Air moisture (characterized, for instance, by the saturation deficit) strongly influences the balance equation through *evaporation*. As a physical phenomenon, evaporation is caused by the escape of oscillating molecules of water from the water surface into the air. If the air is not saturated with water vapour — if there is a saturation deficit d — the water will evaporate, and the greater d is, the greater the evaporation. In addition to d influencing the so-called diffusion evaporation, wind speed also influences evaporation, causing air turbulence and thereby influencing also the exchange of air masses above the surface of water. In other words, when there is a vertical gradient of vapour pressure above any moist surface, a change of state of water from liquid to gas occurs.

In a very thin layer next to the moist surface, diffusivity is entirely molecular. However, the air flow above is not laminar but turbulent. Thus,
a large component of turbulent diffusivity has to be added. Evaporation
can therefore be generally expressed by Dalton's law

$$E_v = \frac{kd}{b},\qquad (2.4)$$

in which E_v represents the evaporation per unit of time (intensity of
            evaporation),
   k = coefficient of wind speed,
   $d = E - e$ = saturation deficit,
   b = coefficient dependent on barometric pressure.

Water evaporates from water surfaces, from the soil and from plants
(transpiration). Each of these types of evaporation depends on many
factors: on the size of the evaporation surface, on its condition and its
physical character, on the type and conditions of the plants and soils,
on the amount of water which can evaporate, etc. All three types of
evaporation are known as *total evaporation*. Evaporation from the plants
(transpiration) and from soil covered by them is known as *evapotranspiration*. The evapotranspiration can be either *potential* or *actual*. Potential
evapotranspiration is sometimes defined as the "water loss which will
occur if at no time there is a deficiency of water in the soil for the use of
vegetation". Under certain conditions, potential evapotranspiration may
be equal to free water evaporation. Actual evapotranspiration is largely
influenced by the availability of water to be evaporated. Thus, significant
differences may exist between potential and actual evapotranspiration,
particularly in arid zones.

Annual total evapotranspiration in terms of catchment can be defined
and calculated as the difference between annual precipitation and annual
runoff, if moisture storage is either small or about the same at the begin-
ing and end of the water year. This method of evaporation estimate,
called the water-balance method, is not applicable for shorter periods,
unless storage data are available (*see page 210*).

In short periods, it is therefore very complicated to calculate the total
evaporation. For evaporation from a water surface, we can use direct
measuring devices called evaporation pans. (The instrument has, however,

a very small evaporation surface and we are extrapolating to a large surface from it.) By this method (*see section 3.6*) we do not measure the actual evaporation but the so-called *pan evaporation* which can be transformed into lake evaporation by pan coefficients. For evapotranspiration, this procedure is less satisfactory. Instruments measuring evapotranspiration are soil evaporimeters or *lysimeters* (*see section 3.6*). Extrapolations from such instruments to larger surfaces or to entire basins are very difficult and the problem of areal evapotranspiration estimation is still at the research stage.

If the air cools below the dew point, part of the vapour contained in it condenses around the so-called condensation nuclei contained in the air. The drops or crystals of water or ice join together in larger drops or crystals and fall to the ground. This process produces *vertical precipitation;* rain, snow, hail, sleet. If the water vapour condenses near the ground or directly on its surface, *horizontal precipitation* results: fog, dew, hoar frost, ground ice. These latter forms of precipitation are not measured by precipitation gauges currently in use.

Precipitation (hydrometeors) is classified also as *liquid* (rain, dew, fog) or *solid* (snow, hail, hoar frost, sleet).

Rain and snow are the most important forms of precipitation for hydrological calculations and balances.

Rain is characterized by the quantity of water which falls, that is, by the precipitation depth hP [mm] or P [m³/km²], and the duration t [min, or h]. In central Europe, the units of volume of precipitation over an area are often expressed in litres per hectare

$$10 \ \text{l/ha} = 1 \ \text{m}^3/\text{km}^2.$$

Intensity i is another characteristic of rainfall. It is a most important characteristic for storm rainfall and it is the ratio of depth of rainfall to the time during which it fell. It is either an average for the entire rainstorm, expressed as

$$i = \frac{hP}{t} \qquad [\text{mm/min}], \tag{2.5}$$

or

$$i_0 = \frac{P}{t} \qquad [\text{m}^3/\text{s km}^2] \text{ or } [\text{l/s ha}], \tag{2.6}$$

Instantaneous intensity (at each moment during the rain) is expressed by the equation

$$\text{instantaneous } i = \frac{\Delta P}{\Delta t},\qquad (2.7)$$

where ΔP is the increase in precipitation during a very short period of time Δt.

Precipitation patterns can be divided, from the meteorological aspect, into two different types: cyclonic precipitation and convective storm rainfall.

Cyclonic precipitations affect a wide area, often entire continents, and result from frontal disturbances during the movement of a barometric low. They cause floods on large rivers if they last very long, even though their intensity may be very low.

Convective storm rainfall is caused either by thermal or orographic convection and affects a relatively small area; it is of short duration but usually of high intensity.

Such rainfalls cause floods on smaller rivers and also in municipal sewers. In addition to their precipitation depth and duration, we also compute their frequency of occurrence (probability). The frequency is ascertained for rainfalls lasting from 5 to 180 min, and for 6, 12, 24, 48, and 72 h. For the most part, storm rains occur in Europe during the vegetation season.

Tropical cyclones and monsoons, although not similar in meteorological synoptic features, have a hydrological aspect almost identical to temperate-zone storm rainfall. This has been corroborated by the author by comparison of hydrologic analysis of central European rains with those of Caribbean hurricanes. Obviously the quantities of rainfall will, however, differ substantially.

The following relation may be established between the average intensity of storm rainfall and its duration

$$i = \frac{A}{t^n},\qquad (2.8)$$

in which i = the average intensity of the rain or its part,

t = duration of the rain or its part,

A = a constant for a given rain gauge station (region) and for a given frequency of occurrence,

n = exponent depending on climatic conditions.

The intensity computed by this and similar formulae is valid for the centre of the rain; the intensity decreases toward the edges of the area affected. According to *Reinhold,* the average intensity on an area of about 10 km² is around 5 per cent and on an area of 25 km² around 10 per cent less than the point intensity calculated from equation 2.8. The relationship is usually satisfactory for durations from 5 to 180 min. For rainstorms over 6 h (12, 24, 48, 72), a depth-area-duration analysis based on isohyetal maps results in depth-area-duration curves. These curves may also be computed with respect to the frequency of occurrence (*see section 4.5d*).

For snow, it is impossible to determine the amount of water that represents the precipitation depth by direct measurements. The snow must be melted and its *water equivalent* (density) computed by comparison of volumes of water and snow. If this equivalent is known, the snow cover can be converted directly to the precipitation depth [mm]. For this purpose, the depth of snow cover and its water equivalent is not only measured at the observation stations, but over the whole basin, with the help of so-called 'snow courses' (*see sections 3.6 and 4.5*).

Daily measurements of precipitation at a single station are added up to give monthly and annual totals. The average annual (long-term) precipitation for a given station is calculated as a mean from several annual values. Averages can be expressed also for monthly and seasonal values. A series of a minimum of 30 years is required for reliable average, called normal. Annual, seasonal, or monthly precipitation series may also be processed by frequency analysis (*see section 4.3c*). The minimum reliable length of series is around 15 years; thus, a probability can be ascertained for different values of precipitation totals. Such computations are particularly needed for irrigation projects.

Depths of precipitation in individual adjacent stations are closely connected and so, if such depths are marked in their respective geographical locations, lines may be drawn indicating places with an equal amount of precipitation. These lines are called *isohyets.* All such maps indicate that precipitation depth is also a function of elevation. This function is called the *pluviometric gradient*. From an isohyetal map, the average precipitation in a basin can be calculated. If the isohyets are not constructed, the average precipitation in a basin can be ascertained directly from measurements of the stations in the basin (*see section 4.5*).

2.3   Surface runoff, the river bed, and the work of water

Precipitated water which is not absorbed into the soil is retained on the earth's surface (depression storage and detention storage) and, if sufficient slope and depth is attained, it runs off, creating overland flow. The overland flow is gradually concentrated in rivulets and semi-permanent small streams creating a network through which water flows into permanent streams, i.e. rivers. These are also fed by sub-surface and groundwater flow; their independence from direct surface runoff results in their permanent character.

Every river has several characteristic reaches: the upper reaches (source or spring region) where it has a steep slope and where the water usually erodes the river bed and carries away sediment and bedload. In the middle reaches, the water may still erode the channel, but it also deposits sediment. In the lower reaches, the stream usually has the least slope and it is here that it deposits the rest of the sediment which it has been carrying, unless it discharges it into the delta or the sea bay.

Because, in most cases, the mouth of a river can be exactly defined, the length of a river is measured from its mouth to its spring, i.e., upstream.

The density of the stream network may be computed by the ratio of aggregate length of all streams in the basin to its surface area and is given in kilometres per square kilometre.

A river network is either symmetrical, if the tributaries of the main stream are situated symmetrically on both sides, or asymmetrical when most of the tributaries flow in from one side.

The hydrographic network also influences the shape of the basin which ranks among its characteristics. The basin is thus characterized by its size and geometric shape (*see section 4.4*) which may be expressed by its average width $\bar{b}$

$$\bar{b} = \frac{A}{L}, \qquad (2.9)$$

where A is the area of the basin,

L = length of the main stream (most often measured on a map), as well as by the average slope of its surface and of the main stream, by vegetation type and extent, particularly of forests, by types and depths of soil in the basin, and by its geological formation.

Streams are fed by groundwater in various proportions and they may be classified in accordance with this. The Russian hydrologist *M. I. Lvovitch* classifies all rivers of the world in this way. All Czechoslovak streams are of the Oder type, characterized by an equal contribution of surface and groundwater to the flow.

During most of the year, rivers flow within their channels. The channels of mountain streams and torrents are irregular and they do not have a clear-cut bank line. Streams and rivers in the plains have more regular channels and most of them have trapezoidal or flat parabolic cross-sections.

During floods, streams overflow their banks and usually flow in the whole river valley. The regularly inundated part of a valley is termed the *flood plain* and a part of it, the *floodway*, is considered a part of the stream.

Streams (with the exception of mountain torrents) flow in alternating curves or *meanders*. The line which joins those parts of the stream which have the highest water velocity (usually the deepest parts) is known as the *thalweg*. The thalweg moves away from the convex bank and approaches the concave bank. At the point at which it crosses over from one half of the stream bed to the other there is frequently a bar. A regular crossing of the thalweg produces a good ford, while an irregular crossing of the thalweg (leap) forms a bad one, mainly because of its instability.

Discharge Q is the volume of water which passes through a section of the channel during a unit of time. The units of discharge are usually cubic metres per second [m³/s] litres per second [l/s = 1 . s⁻¹] and litres per minute [l/min].

In a section of a channel of an area A (*Fig. 2.2*), water particles move

along streamlines with a certain velocity. The velocity of each particle
is different and the streamlines are imaginary lines within the flow, thus
being oriented in many different di-
rections. Let us assume that all the
velocities are the same and equal to
an average velocity v_a and that the
streamlines in the channel are all
parallel. During a unit of time, the
water particles will then progress
by a length l along the stream (*see*
Fig. 2.2). Their space in the different

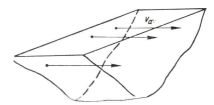

Fig. 2.2. Discharge cross-section

sections which they pass during their motion is immediately filled by
other particles of water so that a compact body of water is formed, the
volume of which represents the discharge Q in the section of an area A.
The volume Q can be calculated as $Q = Av_a$.

The average velocity v_a is thus defined as an imaginary value given
by the ratio

$$v_a = \frac{Q}{A} \qquad [\mathrm{m/s}] \tag{2.10}$$

The movement of water in the channel is basically influenced by three
forces; it is caused by the force of gravity, expressed by the slope and
it is hindered by two other forces, one caused by friction of water against
the channel bed, the other by internal cohesion of particles (molecules)
of water between them — a property of all liquids, known as viscosity.

The flow in an open channel is called steady if the velocity at a point,
or the average velocity in a section, remains constant with respect to
time. If the velocity changes in time, the flow is called unsteady. The
mathematical analysis and solution of unsteady flow are rather complicated and sometimes impossible. Although many phenomena in hydrology,
such as flood waves, involve unsteady flow, their analysis and solution
is transformed so that the principles of steady flow may be applied.

If the velocity does not change with respect to the position of a point
or section, the flow is *uniform*. Conversely, when the velocity changes
with position, the flow is *non-uniform*; this is the case of the flow across
an obstacle in the channel. A steady, uniform flow is rare in nature;
it exists for example, in irrigation channels which have a regular cross-

section and a constant rate of discharge. For practical purposes, however, a flow is considered uniform if the depth in the direction of flow remains constant. The flow in an open channel may be also laminar or turbulent. In a laminar flow, the streamlines remain parallel to each other; conversely, in the turbulent flow, they are mixed. Finally, an open channel flow is either *tranquil* (sub-critical) or *rapid* (super-critical). The distinction is dependent on the depth at which a certain discharge passes through

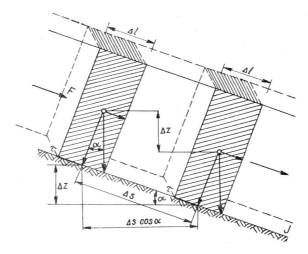

Fig. 2.3. Open channel flow

the channel. Two such depths are possible: the higher one corresponds to the tranquil flow, the lower one to the rapid flow. The boundary between such depths is called the critical depth; at such depth the flow is called critical.

The water in a river channel is in contact with the river bed and with the air above the surface. The part of a channel cross-section in contact with the water is called the wetted perimeter P. The ratio of the area of the cross-section A and of the wetted perimeter P is called the hydraulic radius R, thus

$$R = \frac{A}{P} \tag{2.11}$$

The water in its movement is working on the shape of the channel. By evaluating the work of the water, the average velocity of water in a uniform steady flow may be ascertained.

If an open channel is assumed, with a constant slope I and a constant depth h, the water surface slope will be parallel to the channel slope. The weight of an element of the stream (*see Fig. 2.3*) of an area A and of a length Δl represents a force G, which may be resolved into its components f_1 perpendicular to the bottom and f_2 parallel to it (and thus to the slope). These forces are

$$G = \gamma A\ \Delta l,$$
$$f_1 = G \cos \alpha,$$
$$f_2 = G \sin \alpha,$$

where γ = specific weight of water and α = angle of the slope.

The effect of a constant action of gravity is an accelerated motion (for example, a free fall). If the water motion in the channel is uniform, as stated above, other forces are acting in the channel, as already mentioned. There is, in the first place, the friction of water against the channel (the drag). This force is equal and acting in the opposite direction to the force $f_2 = G \sin \alpha = \gamma A\ \Delta l \sin \alpha$. The friction force F is directly proportional to unit friction τ (on a unit of surface) or the drag, and to the area subjected to friction, i.e., the part of the channel in contact with water. Thus

$$F = \tau P\ \Delta l. \tag{2.12}$$

Further, $F = f_2$ as stated above, and so

$$\gamma A\ \Delta l \sin \alpha = \tau P\ \Delta l$$

or

$$\frac{A}{P} \sin \alpha = \frac{\tau}{\gamma}.$$

The angle α is generally relatively small and $\sin \alpha$ can be replaced by $\tan \alpha = I$; further $\dfrac{A}{P} = R$ (hydraulic radius). Thus

$$\frac{\tau}{\gamma} = RI.$$

The ratio of unit friction τ and the specific weight of water γ expresses the loss of hydraulic head of flowing water which is directly proportional to the square of the average water velocity or

$$\frac{\tau}{\gamma} = nv_a^2.$$

where n = coefficient of proportionality, mainly dependent on the roughness of the channel.

Thus

$$RI = nv_a^2,$$

$$v = \sqrt{\frac{1}{n} RI} = \sqrt{\frac{1}{n}} \sqrt{RI}.$$

The coefficient $\sqrt{\dfrac{1}{n}}$ may be written as C; thus

$$v = C\sqrt{RI}.$$

This equation, proposed by *Chézy* in 1769, is still the basis of many hydrologic computations of discharges. The *Chézy* discharge coefficient C includes the roughness coefficient n, which is named after *Manning*, who proposed it in 1889.

The unit friction τ or the drag can be expressed from the above equations as

$$\tau = \gamma RI.$$

If the drag is larger than the cohesive forces of the material forming the channel, this material will be eroded and carried away by the water. By assuming a relatively small depth of the channel in respect to its width, the hydraulic radius R may be replaced by the depth of the flow h. Thus the drag force can be expressed as a unit *tractive* force

$$\tau_0 = \gamma hI \qquad [\text{kg/m}^2].$$

For erodible channels which scour but do not silt, this equation is used for computation of sediment transport.

Sediment transport plays an important part in hydrological computations. Two types of particles are transported:

(a) The suspended sediment.

(b) The bed load, which is dragged in constant contact with the bed or by saltation (jumps). The suspended sediment is of a finer size, and its content in the water is expressed by water turbidity [kg/m^3, g/m^3, or mg/l].

The unit tractive force permits the computation in some simple cases of the *permissible velocity* which will not cause the erosion of the channel bed. This velocity is, however, dependent on many other factors and is consequently fairly unstable. Other methods are being developed at present to compute the elements of erosion of beds, a problem that is not yet solved to the satisfaction of the hydrology scientist and practising engineer (*see Chow, 1964, section 17*).

2.4 Factors influencing the surface runoff and characteristics of streamflow

The regime of a stream is the result of the runoff process in its basin, and so the characteristics of the basin will be the main factors of runoff. They may be divided into two main groups: geographical factors and geomorphological factors.

(a) Geographical factors

Climatic (hydrometeorological) factors. The most important are precipitation and total evaporation (evapotranspiration) and of less importance are the air temperature, humidity (saturation deficit), air pressure and the direction and speed of wind. While precipitation and evaporation have a direct influence on the water balance in a basin, and hence have an influence both on the total volume of surface and sub-surface runoff and flow variations, the other climatic influences, such as air humidity, temperature or pressure and wind have an indirect influence on runoff and are expressed finally again as variations in precipitation and evaporation.

Soil and geological conditions in the basin. These are of major importance in the distribution of runoff between surface and sub-surface runoff.

Permeable soils increase the sub-surface runoff at the expense of surface runoff, slowing down the total runoff, and thereby exerting a favourable influence on the equilibrium of the flow in the stream during the entire year. Impermeable soils have the opposite effect.

The geological substrata of the basin expresses itself in much the same way, but its influence is mainly in the larger basins, while soil conditions are more influential on smaller basins. As far as peat is concerned, it should be pointed out that it is erroneous to consider it as a reservoir of groundwater. Peat does absorb water well, but it releases it into the runoff only with difficulty; hence, it rather evaporates, with a negative influence on the hydrological balance. Peat which is fully saturated with water acts in the same way as impermeable soil and so the surface runoff from it is rapid, and thus unfavourable.

Vegetation cover may act favourably or unfavourably on surface runoff. Forests are the most important vegetation factor influencing surface runoff and groundwater flow. Yet only forests of a proper composition and with the right kind of location in relation to the watershed exert a favourable influence. From the hydrological standpoint, the best type is a mixed forest with large quantities of litter located on the watershed divide line, or near it. A great deal of transpiration from forests is, however, unfavourable for the hydrological balance. Meadows protect the soil against erosion while partially slowing runoff and promoting its absorption.

Artificial and natural reservoirs, even though they increase somewhat the evaporation from water surfaces in the basin, act favourably on the distribution of runoff during the year and regulate flood flows.

Man's activities influence the quantity of runoff, mainly through the land use. Proper land use (contour ploughing, discing) and crop rotation are important in this regard.

(b) Geomorphological factors

Size and shape of basin. The size of a basin has no basic influence on the average long-term volume of its runoff, but it is very important for flood flow and the distribution of flow during the year. The smaller the basin, the greater the flood flow from its unit area and the less balanced the distribution of the runoff.

The shape of the basin also mainly influences the flood flow. The longer and more wand-like its shape, the smaller the flood peak will be, regardless of the size of the basin. Fan-shaped basins have the highest flood peaks.

The character and slope of the main stream and of the basin have considerable influence, both on the average value and on the extreme characteristics of the runoff. The steeper this slope, the greater the surface runoff will be at the expense of groundwater flow, and its distribution over the year will not be balanced. The flow will also be larger.

The number of factors influencing runoff is great: they are inter-connected and often contradictory. At the same time, despite the great progress which has been made in engineering hydrology, the influence of some of these factors has not yet been sufficiently studied.

(c)   Characteristics of surface runoff

It is evident that, with the exception of small experimental plots, surface runoff as overland or sheet flow on the basin cannot be directly measured. Its value can be ascertained from the discharge measured in the main stream at the lowest point of the basin. The discharge is the result of runoff and they are two characteristics of the same process.

The basic measurement of discharge at a stream gauging station is one or several daily observations of water stage (level), or its continuous registration, providing that this station has a stage-discharge relation established. A more detailed consideration of these measurements is given later (see section 4.6).

If the stage does not vary much during a day (which occurs in periods of low flows and for some streams, in arid regions — Tigris and Euphrates, for example — for low flows, even for longer periods, such as weeks) a *daily average discharge* may be computed from a single daily measure-ment. If the variation during the day is appreciable, several daily measure-ments are necessary for an average to be computed, and, whenever possible, a continuous recording stage gauge is used for this purpose.

From the average daily discharges $Q(d)$ during a month, an average monthly discharge $Q(m)$ is computed as an arithmetic mean. Similarly, an average annual discharge $Q(y)$ is ascertained. From several average annual discharges, an arithmetic mean may be computed to give a long-term *mean annual discharge* $MQ(y)$. Similar long-term means can be computed for monthly discharges $MQ(m)$ (for example, a mean monthly discharge for July). Long-term mean daily discharges are seldom computed, since they are physically meaningless.

From these characteristics of discharge Q, the volume of runoff V during a given period can be determined by multiplying the discharge by the number of seconds in the period. For example, the daily volume is ascertained by multiplying $Q(d)$ by 86 400 (number of seconds in a day).

Expressing the characteristics of a basin (or stream) in this way does not permit hydrological comparisons to be made with other basins (streams) because it is obvious that the $MQ(y)$, for instance, from a large arid basin will always be greater than the $MQ(y)$ of a small basin in conditions of sufficient or even excessive moisture. A *specific* or *unit runoff q* which expresses the unit average contribution of the basin to the discharge may be expressed with the same average indices as Q.

$$q = \frac{Q}{A} \quad [\text{m}^3/\text{s km}^2, \text{l/s km}^2, \text{l/s ha}] \quad\quad (2.13)$$

Q = discharge,
A = area of basin.

Thus, the long-term mean annual unit runoff will be $Mq(y)$, the average monthly specific runoff $q(m)$, etc.

Discharge, as well as unit runoff, are momentary values. In view of the fact that, in balance equations, precipitation and evaporation are usually given in depth units for a certain period (hP = precipitation in mm; hE = evaporation in mm), discharge q can also be expressed in runoff depth hR. The conversion factors are as follows:

$$1 \text{ mm of runoff} = 1000 \text{ m}^3/\text{km}^2$$

and

$$hR = \frac{qt}{1000},$$

where hR = runoff depth [mm],

   t = period for which it was computed,

   q = unit runoff [$m^3/s\ km^2$].

The ratio of the runoff depth hR to the precipitation depth hP for a given period is called the runoff coefficient C

$$\frac{hR}{hP} = C.$$

The above values characterize the averages of the regime of streams. They give no information on the fluctuations in the regime from year to year or on storm runoff and distribution of flood flow.

For a simple characterization of the annual yield of the basin, coefficients of average annual flows $Q(y)$ may be computed as their ratio to the long-term average annual flow $MQ(y)$.

$$k_i = \frac{Q(y)_i}{MQ(y)}.$$

If $k_i > 1$, it is a year with an excess of moisture: if $k_i < 1$, it is a dry year.

A more precise characterization, useful for economics of water resources, is the frequency analysis of annual and monthly yields of the basin, or mean discharges. A probability of occurrence for different values of these yields may be computed either by plotting the data on a probability paper or by curve fitting and subsequent extrapolation of recurrence interval.

The fluctuation of flow during a calendar year is expressed by the sequence of average monthly flows $MQ(m)$ for each month of the year.

The distribution of flow during the year according to availability of volume of flow (not calendar distribution) is characterized by a duration curve of average daily flows $Q(d)$. It can be plotted for one or more years, as will be seen later.

On the basis of these curves, the duration of different magnitudes of flow during n days is given by the n-daily flows Q_n. Thus, for example, a very low flow which, according to central European standards, must always be left in the stream while withdrawing water from the stream, is a 355-day flow (Q_{355}), i.e., a flow which is reached or exceeded 355 days in a year and which is not reached only during 10 days in the year. On the other hand, a 30-day flow (Q_{30}) is a high flow, because it was reached or exceeded only during 30 days of the year, while during the rest of the year, i.e., for 335 days (about 11 months), the flow was lower.

In streams with a regular regime, the flow fluctuations during the year (such as on some rivers in the USSR, where the spring runoff of snow water and low flow in autumn are relatively regular) can be expressed by a standardized hydrograph computed from the annual hydrographs.

For most of the projects in a basin and on a stream, the extreme values of flow, i.e., flood flow and low flow, are often more important than averages.

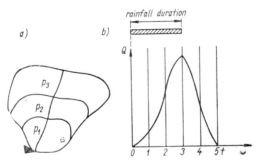

Fig. 2.4. Hydrograph formation

A *flood* was formerly characterized as a catastrophic flow. But one cannot say unequivocally that all floods are disastrous. Some spring floods can be damaging in some reaches of the stream, for instance, where they flow through settlements, flooding buildings and disrupting communications. But this same flood may inundate meadows, leaving behind fertilizing deposits which can be beneficial rather than destructive.

This is why the probability of occurrence of floods has been introduced in engineering hydrology and permits an economic evaluation and classification of peaks and volumes of floods.

In this way the flood peak flow or volume with a recurrence interval of 1, 2, 5, 10, 20, 50, or 100 years is called an *N*-year flood, which can be indicated by the symbol HQ_N or in the probability percentage HQ_p. The symbol of the highest peak discharge which has occurred during the period under observation may be HHQ (period), while the average peak discharge (arithmetic average from all observed peak discharges) is MHQ (period).

In most cases, in the computation of floods, it is the peak discharge, the flood volume, and sometimes the shape of the hydrograph of the flood in a given section which are required. It is less often necessary to determine the real shape of the flood wave along the stream and the speed of its advance. These last indications are, however, most important for real-time hydrological forecasting.

The process of runoff in a basin during a flood highlights all the salient phases of the runoff process, and that within a short period of a few hours or days. That is why the analysis of floods and of the flood hydrograph is one of the most important problems of engineering hydrology. The mechanism of the flood hydrograph formation can be summed up as follows.

If P is the surface of a given basin (*Fig. 2.4a*), entirely covered by rain, the process of runoff from a rain which affected the area of the entire basin may be analysed, it being assumed for reasons of simplicity that the rain intensity is constant in time and space. The basin may be divided into smaller parts p_1, p_2, p_3, or in general p_i, in such a way that water travels from each part to the lowest point of the basin in the same time t_1, t_2, t_3, or in general t_k. These parts are called *isochrones*. Only some of the rain which falls on each isochrone surface flows off.

The isochrone is, in its turn, composed of many elementary runoff plots and on each of these a basic hydrological balance may be expressed

$$q_0 = i_0 - z_0, \tag{2.14}$$

where q_0 = elementary surface runoff,

$\quad\quad i_0$ = intensity of rain,

$\quad\quad z_0$ = losses.

As a rule, losses of surface runoff may occur by infiltration and evaporation. Whereas infiltration losses occur quickly, almost exclusively during the rain, evaporation (evapotranspiration) is a continuous process and its rate is negligible with respect to infiltration in periods of flood. The above equation can thus be written in the form

$$q_0 = i_0 - f_0,$$

where f_0 = infiltration on the elementary runoff plot.

From this equation, it appears that the condition for the existence of surface runoff is that $i_0 > f_0$, or that the rain intensity is greater than the infiltration rate. The runoff from elementary runoff plots is therefore also called 'rainfall excess'.

The runoff from the entire isochrone surface will be an average in time and space of rainfall excess on its elementary plots q_i. The volume of runoff from the isochrone surface during an interval of time t_k will be

$$V_i = q_i P_i t_k$$

and the discharge during this interval

$$Q_i = q_i P_i.$$

The analysis in time of the discharge concentrated and arriving at the lowest point of the basin, at a cross-section of the stream, will show the composition of this discharge as follows.

During the first interval t_1, the runoff from the first isochrone will arrive at the section and the discharge will be

$$Q_1 = q_1 P_1. \tag{2.15}$$

During the second interval t_2, the runoff from the first interval of the second isochrone will also arrive at the section and therefore

$$Q_2 = q_2 P_1 + q_1 P_2.$$

And during the third interval, similarly (*see also* Fig. 2.4)

$$Q_3 = q_3 P_1 + q_2 P_2 + q_1 P_3.$$

At this moment, the entire basin is contributing to the runoff and therefore the flood peak is attained—a state of equilibrium on the basin is reached. The time needed to reach this moment is sometimes called the *concentration time* and it is identical with the *critical rain duration*. This process also forms the rising limb of the hydrograph. If the rain continues, the peak of the hydrograph (or flood) is stationary. When the rain ceases, each isochrone in turn ceases to contribute to the discharge and the falling limb of the hydrograph is formed. In practice, this falling limb is much less steep than the rising one, mainly because of the storage in the basin.

Experimental work with physical models of basins conducted by the author (Němec, 1968) fully confirms this simple analysis. If the concentration time is T_c and the rain duration T_k, the time of direct runoff from the basin will be $T_c + T_k$.

Often, in a so-called rational formula for flood peak estimation, the maximum constant rain intensity is reduced to the runoff by a coefficient of runoff C_m

$$HQ = i_m C_m A. \tag{2.16}$$

It is most important not to confuse this coefficient of runoff, which is without any physical sense, with the coefficient of volume of runoff, for example, on elementary runoff plots, which expresses the value of infiltration by the ratio of volume of surface runoff to volume of rainfall.

This theoretical analysis and synthesis of the flood hydrograph (*see page 239*) is based on many simplifications and assumptions and serves the purpose of explaining the runoff process and the concept of the hydrograph. The use of simplifications is nevertheless often justified, because without them the estimation of flood flow without a long-term series of direct measurements of discharges would not only be very difficult, but often even impossible, and engineers cannot get along without flood peak data estimation.

If the flood wave advances along the stream or overflows the banks and flows in the flood plain, it changes its shape and the peak discharge is generally diminished. This transformation of the flood wave, or so-called flood routing, is caused in particular by reservoirs or by flood-plain storage. It should be pointed out that flood routing does not concern the volume of water which is stored (retained) in the reservoir. It is the storage in the flowing mass of water that is considered by the flood routing procedures, either when this mass travels through a reservoir or through open channel reaches (*see section 5.3*).

Reservoir detention and flood-plain routing were often computed graphically; today, extensive use is made of analogue and digital electronic computers. Nevertheless, as a rule, the flood hydrograph computation and characteristics of the reservoir (flood-plain), such as area and volume curves with respect to the water depth in the reservoir, are prerequisites for reservoir design.

Minimum discharge, or low flow, is the result of a decline in, or complete absence of, surface runoff during dry periods. During such periods, surface streams are supplied for the most part by groundwater, on the quantity of which depends the size of the lowest flow. In the mountain streams of Czechoslovakia, minimum discharges begin at the end of winter and, on other streams, especially in the plains, at the end of a dry summer.

The following features characterize minimum discharge:

(a) NNQ, which is the smallest average daily discharge for the entire period under observation;

(b) $MNQ(y)$, which is the arithmetic average of the lowest annual discharges, on the basis of which it is possible to classify the individual annual minima by means of a ratio coefficient or frequency (probability), such as $NQ_{p\%}$ or NQ_N;

(c) a characteristic most commonly used in central Europe is Q_{355} and Q_{364}, i.e., mean daily discharges which were exceeded during 355 or 364 days, computed from the duration curve of the mean daily discharges of an average year whose $MQ(y)$ has a probability of 50 per cent.

Minimum discharges take on greater importance with the growing demands for water and the increasing pollution of streams. Methods for determining and observing them have not yet been sufficiently developed, however.

All the characteristics of discharges in a stream are applicable also for water stages (height of water level) which are connected to the discharges. *Table 2* reviews all the hydrological characteristics with their symbols as used in central European standards (DIN, ÖN, etc.).

Ice forms on many streams during the winter. If the water temperature drops below 0 °C, ice begins to form, first on the water surface along the banks where the current is least rapid. Narrow but connected bands of bank ice are thereby formed along the stream banks. The current breaks up the ice and carries it away. If the current is swift and if the cold weather lasts for some time, chunks of porous ice form in the middle

Table 2. Symbols for statistical characteristics of hydrological elements in analogy with DIN, ÖN, and other standards

Symbol	Characteristic	Remarks
Q (d)	average daily discharge	
Q (m)	average monthly discharge	average of daily averages Q in this month
Q (y)	average annual discharge	average of daily or monthly averages in this year
MQ (y)	mean annual discharge	long-term mean of several averages of annual Q
MQ (m)	mean monthly discharge	long-term mean of several average monthly Q for a particular month (e.g., July)
$RHHQ$	maximum probable flood peak discharge	theoretically computed maximum peak discharge
HHQ	maximum observed flood peak discharge	for all the period of observation
HQ	maximum peak discharge	during a certain period
MHQ	average maximum peak discharge	average of all peaks during a certain period
$MYHQ$	average of annual maximum peak discharges	
NQ	the lowest discharge during a certain period	
NNQ	the lowest discharge ever observed	
HQ_N	N-year peak discharge	a maximum discharge of an N-year recurrence interval
$HQ_{p\%}$	$p\%$ probability peak discharge	a maximum discharge of $p\%$ probability of occurrence
Q_n	average daily discharge of a duration of n-days (for one particular year)	computed from a duration curve of daily average discharges for one year
MQ_n	daily average discharge of a duration of n-days (for a long period)	computed as above from a duration curve including a series of years

of the river — *frazil ice*. This, together with the pieces of ice from along the banks, can combine to form *sheet ice* with rounded edges which can in turn join to form a solid ice cover with an uneven surface at bends of the stream or at narrow places. The sheet ice is pushed under the surface, the stream cross-section is blocked and an *ice jam* results. In slower streams, the bank ice on both sides extends into the middle of the stream until it freezes over and a smooth ice cover results. This ice cover may or may not follow the fluctuation of the water stage. In the spring, the ice melts, with warmer weather and increased discharges, and ice jams result, as a rule in bends and narrow parts of the stream.

A porous, spongy *anchor ice* is formed at the bottom of some streams, especially in the mountains, although the surface is ice-free. It does not freeze over, as the current is very swift. This type of ice may even form thick layers which cut down the cross-section of the stream bed, thereby raising the water stage without a correspondingly larger discharge.

2.5   Sub-surface water

Sub-surface water may be basically classified into soil water (soil moisture) and groundwater.

(a)   Groundwater

Groundwater originates from surface water. The infiltration through soil reaches a zone where all the pores are saturated; the water flow and movement is then controlled by hydraulic boundary conditions. Thus, a basic distinction exists between the soil water, which occurs mostly in the unsaturated zone, and groundwater occurring in the zone of saturation or aquifer. An aquifer is a permeable geological stratum in which there is a storage of groundwater. There are laws of groundwater movement into, through, and out of an aquifer. This movement can be either natural or artificial, the latter being the result of human activities, such as artificial recharge or groundwater extraction by wells. The basic law of groundwater motion is *Darcy's law*

$$v_g = kI,\qquad\qquad (2.17)$$

where v_g = velocity of groundwater motion,
$\quad\quad I$ = hydraulic head (or gradient),
$\quad\quad k$ = *Darcy's coefficient of proportionality* — indicating the permeability (or the hydraulic conductivity) of the porous medium.

Any aquifer is conditioned by an impermeable stratum, confining bed or ceiling, called an aquiclude. Aquifers and aquicludes can be stratified so as to form several horizons of groundwater. If the groundwater level is free, an unconfined aquifer of phreatic water results; this is the case with most aquifers attained by shallower wells. If the groundwater is confined by an aquiclude ceiling, an aquifer under hydrostatic head (pressure), or artesian groundwater results.

Analysis of groundwater motion and its utilization for well hydraulics, groundwater fluctuation computations and other particular aspects of groundwater movement, such as artificial recharge and salt-water intrusion in aquifers (near coasts), are the subject of geohydrology and are not treated in this book.

(b) Soil water

Soil water includes all water in the pores between the solid particles of soil. The following types of soil water may be ascertained according to the size of the pores and the forces acting at the interface of the solid and liquid phase.

In the sub-capillary pores the molecular adsorbant forces act through the whole pore and the water retained there (in addition to water which is chemically incorporated in the grains of minerals) is the hygroscopic water (pellicular, funicular).

Pellicular water is composed of one or several layers of molecules around a grain. Funicular water is detained in the corners of contact between two grains.

Quasi-stationary capillary water occurs here too; it is called pendular water. This water is stationary since the adsorbant forces which act on it amount to tens of thousands of atmospheres. The magnitude of sub-capillary pores is between 0.1 and 3.0 μ. Water which is retained in them is inaccessible to plants.

Most of the soil water in the capillary pores is under the influence of capillary molecular forces of surface tension, creating menisci on the surface of the water; this can move in all directions, regardless of gravity. Capillary water can either rise as a capillary fringe of the groundwater, when it is *sustained capillary water*, or it can infiltrate from the soil surface, not connecting with the groundwater, when it is called *suspended capillary water*.

Capillary pores have a dimension of up to 0.05 mm; the forces acting on capillary water are relatively small — between one-tenth and one-hundredth of an atmosphere. This water is thus freely accessible to plants, and is the main source of soil moisture used by plants and, unfortunately, for evaporation from soil also.

In non-capillary soil pores, water moves under the influence of gravity. It is called *gravitational water*. Yet even in these pores, water is detained at the interface by strong molecular forces; adsorption forces also act in the capillary pores and hold all the types of water described in connection with sub-capillary pores.

Depending on the amount of different types of water in the soil, the soil attains different stages of saturation, separated at points called *soil hydrological limits* (*hydroconstants* or *hydrolimits*).

At maximum water capacity, all pores are filled with water. At field-water capacity, gravitational water is drained from the soil profile and it is mainly the free capillary water which is retained in the capillary pores. At wilting point, there is no water in the soil accessible to plants. These three limits are the most important for the hydrologist.

The numerical value of the soil hydrolimits varies within a range not only for different soils, but also with respect to a period or a season of the year. This range of moisture depends mainly on the texture of the soil and on its porosity.

From the standpoint of hydrological calculations, the field-water capacity or its equivalents are of particular importance. A review of soil-water limits and capacities, as ascertained by Czechoslovak soil scientists, is included in *Table 3*.

The movement of soil water is important for the hydrologist, in particular, with respect to infiltration, which plays a primary role in the analysis of surface runoff process and flood hydrograph synthesis (*see section 4.8*).

Table 3. Hydrological soil moisture limits (hydrolimits)

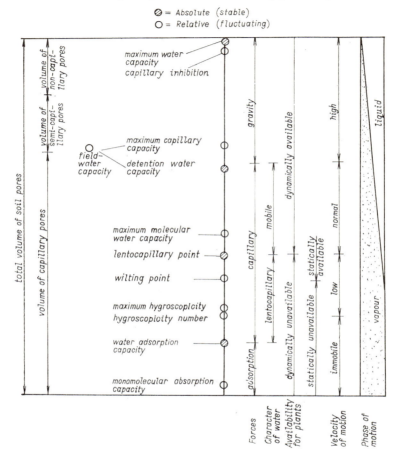

2.6   Conclusion

This short review of basic principles governing the different phases of the hydrological cycle is intended to put into the right perspective the practical part of the book which follows.

A more detailed physical and mathematical description of these principles is included in references, to which the reader is directed.

3 Hydrological and Hydrometeorological Observations, Measurements and Instruments

3.1 Organization of hydrological and hydrometeorological services

Without hydrometeorological and hydrological observations and measurements, the principles described in chapter 2 could not be derived. However, the difference between engineering hydrology and geographical hydrology is that the second deals with the qualitative descriptions and relations of phenomena, while the first is based on engineering requirements and derives numerical data for them. It could be said that the qualitative nature of a phenomenon is ascertained by observing it, and its quantitative nature by measuring it. In view of the fact that, in engineering hydrology, observations and measurements are for the most part inter-connected and complementary, both terms are considered as equivalent. Hydrological and hydrometeorological phenomena are observed and measured at special stations, usually by volunteers. These observers are given simple, unified instructions and their work is checked by professional staff. The network of all observation stations for meteorology and hydrology is administered in Czechoslovakia by a unified service: the *Hydrometeorological Institute*. Its climatology divisions look after hydrometeorological (climatological) observations, while its hydrology division is responsible for hydrological measurements.

A recent survey by the *World Meteorological Organization* shows that in 23 countries in different parts of the world, such as the USSR, Sweden, Guyana or Cambodia, there exists a joint national service administering both meteorological and hydrological networks. In 42 countries, these services are closely co-ordinated. In 49 countries, the meteorological and hydrological services, and thus also the networks, are under separate administrations. While there is, as a rule, only one central meteorological authority in these countries, many of them have several hydrological agencies which provide measurements of hydrological elements, such as stream discharges.

In almost all countries, groundwater observations and, in particular, groundwater prospecting are under a different administration from the meteorological and surface water networks.

3.2 Hydrometeorological (climatic) observations and data

Almost all the countries of the world (131) have accepted the World Meteorological Organization's *Technical Regulations* (1968) on meteorological observing stations, their networks, observing practices and procedures, and processing and publishing of data. These regulations secure uniformity and standardization. In addition, WMO has published a series of guides and manuals which contain descriptions of recommended practices and instruments.

According to the WMO Technical Regulations, accepted in Czechoslovakia, meteorological observing stations are classified as:

(a) synoptic stations;
(b) climatological stations;
(c) agricultural stations;
(d) aeronautical meteorological stations;
(e) special stations.

For hydrological purposes, data from climatological, agricultural meteorological, and special stations are of predominant importance.

Climatological stations are principal, ordinary, precipitation, or special-purpose stations.

At a principal climatological station, observations are made of all or most of the following elements: weather, wind, cloud conditions, temperature (including extreme temperatures), humidity, atmospheric pressure, precipitation, snow cover, sunshine, soil temperature. Soil temperature should be measured at some or all of the following depths: 5, 10, 20, 50, 100, 150, and 300 cm.

At an ordinary climatological station, observations are made of extreme temperature and amount of precipitation.

A precipitation station measures precipitation either daily or continuously.

Agricultural meteorological stations conduct detailed observations of physical environment and observations of a biological nature. In addition to precipitation, air temperature and humidity, such stations measure soil temperature, soil moisture, evaporation — from water surfaces as well as from the soil and plants — and turbulence and speed of wind in characteristic layers of air near the surface of the ground. They also observe phenological data (plant growth in connection with weather developments), the spread of plant diseases, etc.

Special stations serve purposes not covered by other stations and include in particular stations for hydrology and for radar detection of precipitations, for microclimatology, and for measurement of radiation. These latter are most important for evaporation (evapotranspiration) estimations.

Stations are set up so that the result of their observations is representative, i.e., reflects meteorological conditions not only at the station itself but also in a wider area around it. The network of stations should be sufficiently dense and they should be situated evenly throughout the whole territory and, insofar as possible, at different elevations.

3.3   Optimum network of climatic stations for hydrological purposes

It is useful at this point to consider an optimum density for climatic station networks. As a result of international effort to establish norms of network density, the WMO Guide to Hydrometeorological Practices (1970) divides the areas of different density of stations into three regions, as follows.

(*a*)   *Flat regions of temperate, mediterranean, and tropical zones*

11 − 17 stations for 10 000 km^2; 1 station for 600 − 900 km^2.

But for countries where it does not seem possible to achieve, for the time being, the density required, because of the low density of the population, the lack of development of communication facilities, or for other economic reasons, a reduction of density of the raingauge network is accepted to the range of:

1 station for $900-3000$ km^2.

(b)   Mountainous regions of temperate, mediterranean, and tropical zones

In mountainous regions, it is most desirable to have stations distributed in altitude zones of approximately 500 m per zone and with minimum density:

$40-100$ stations for 10 000 km^2; 1 station for $100-250$ km^2.

But for such countries where it does not seem possible to achieve this density for the time being, for the same reasons as those mentioned for category (a), a density of the raingauge network should be attained of:

1 station for $250-1000$ km^2.*

On the other hand, for small islands less than 20 000 km^2, with very irregular regimen and very dense stream network, the minimum density should be:

400 stations for 10 000 km^2; 1 station for 25 km^2.

(c)   Arid and polar zones

$1-7$ stations for 10 000 km^2; 1 station for $1500-10\ 000$ km^2 depending on the feasibility.

These are not applicable to the great deserts with no organized hydrographic network (Sahara, Gobi, Arabian, and so on) and great ice fields (Antarctic, Greenland, Arctic islands). In these regions, precipitation is not studied by raingauge networks of standard type but by special stations and methods of observation.

Some regions will fit into one of the above categories while others will not. Some regions, e.g., hilly regions, may be such that they will need a density, say between (a) and (b).

In all cases, where the minimum is fixed at less than 1 station per 1000 km^2 for the regions in categories (a) and (b), selected small areas of the greater density given for that category in order that representative information may be available on the variability of precipitation.

---

* In very difficult conditions, this may be extended to some 2000 km^2.

3.4   Instruments in minimum networks

The minimum network should consist of three kinds of gauges:

Standard gauges. These gauges are read daily. Besides daily depth of precipitation, observations of snowfall, depth of snow on the ground, and state of weather are to be made at each standard precipitation station.

Recorders. It is desirable to aim at having at least 10 per cent of the stations equipped with recorders in warm climates, and 5 per cent in cold climates.

Storage gauges (*totalizers*). In sparsely settled or remote regions, such as in desert or mountainous terrain, storage gauges may be used. These gauges are read monthly, seasonally, or when it is possible to inspect the stations.

Location of precipitation gauges relative to stream-gauging network. In order to ensure that precipitation gauges are available for extending streamflow records, for forecasting purposes, or for water-balance analysis, the co-ordination of precipitation gauges with the stream-gauging network should not be left to chance. Precipitation gauges should be located so that there are at least two precipitation gauges for each stream-gauging station: the following procedures would be helpful.

(a) Install a standard gauge at all the river-gauging stations.

(b) Install a second raingauge in the upper part of the small river basin above the river-gauging station.

It is highly useful to supplement the formal precipitation network by observations of rainfall caught in exposed vessels after the occurrence of intense rainstorms. But these bucket surveys are not to be considered as part of the minimum network.

From the hydrological standpoint, climatological and agricultural meteorological stations are particularly important for providing data on precipitation, air temperature and humidity (saturation deficit), wind speed, and, in some cases, direct measurements of evaporation.

It is therefore in the interests of everyone who works with hydrological data that the network of climatological (meteorological) stations be as dense as possible. It would be only proper if engineers themselves would show interest in setting up such stations in smaller basins connected with their projects. It is not difficult or expensive to set up such stations.

3.5 Climatological stations for hydrological purposes

The basic condition for the installation of a good climatological station is the assurance that observations will be continued consistently according to the regulations of the hydrometeorological service. These regulations, based on an international agreement, request that each climatological station be located at a place and under an arrangement which will provide for the continued operation of the station for at least 10 years, and for the exposure to remain unchanged over a long period unless it serves a special purpose which justifies its functioning for a shorter period.

Stations should be located in open space typical of the area which they are to represent. The location of the raingauge must be chosen so that it is not near to any other structure or object (tree, house) which is more than twice its height (*see also section 3.6g*).

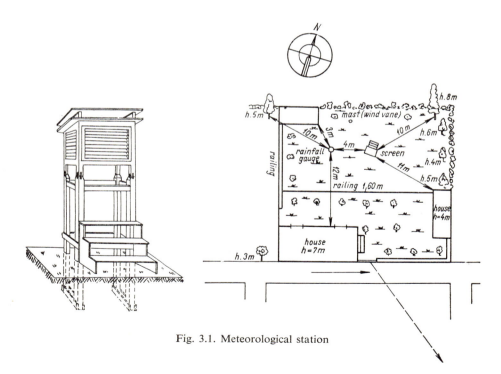

Fig. 3.1. Meteorological station

The size of the station area varies, depending on the number of instruments (between 6×9 and 25×25 m). It should be fenced in.

Instruments are either arranged in the open (raingauge, evaporation pan, wind vane) or in a *meteorological louvred screen* (*Fig. 3.1*) so that air may penetrate but not direct sun rays. The screen has a thick wire net or staggered boards in place of a floor. Its dimensions are generally standardized at national level which, in Czechoslovakia, is $80 \times 60 \times 60$ cm. If the station comprises only a raingauge and thermometer, a small metal shield to shade the thermometer may be sufficient.

In central Europe, the instruments should usually have their sensor element 2 m above the ground. The 'grass' minimum thermometer should be 5 cm above the ground, the raingauge rim 1 m above the ground, the wind vanes and anemometers between 8 and 10 m above the ground (on masts). Height is not important for sunshine recorders but this instrument must not be shaded by any obstacles, particularly from the east or west.

Once installed, the station must not be moved even a short distance, because this would destroy the homogeneity of the observations in space.

In order to assure homogeneity in time, the stations (with the exception of synoptic stations) conduct their observations according to mean local (sun) time, which differs from the zone times. Since the meteorological phenomena have a daily cycle with respect to the sun's position, the observations are made at fixed hours according to local mean time or GMT, which remain unchanged throughout the year. In Czechoslovakia, these fixed hours are 0700, 1400 and 2100 h, local time.

The observer records the readings in the station's log book, from which they are usually transposed into a monthly report, sent to the headquarters of the Service. An example of a monthly return from a station log book is given in Appendix 7. The log book and report contain numerical data as well as weather reports indicated in internationally agreed codes. The reports are first verified and then processed at the headquarters. The processed data are published and used in maps and graphs.

Where national meteorological services have not published their own regulations and guidance for climatological stations, instruments, and methods of observation, the following publications of the World Meteorological Organization, internationally recognized, may be consulted: Technical Regulations, Third Edition, 1968 (Publication No. 48. BD. 2); Guide to Meteorological Instrument and Observing Practices, Third Edition, 1969 (Publication No. 6. TP. 3); Guide to Hydrometeorological Practices, Second Edition, 1970 (Publication No. 168. TP. 82); Guide to Climatological Practices, First Edition, 1960 (Publication No. 100. TP. 44).

If the necessary hydrometeorological (climatic) data are not published in meteorological yearbooks, or are needed in specially processed form, the national meteorological service may supply them on request. In Czechoslovakia, the Climatic Division of the Hydrometeorological Institute provides the following data from the nearest meteorological station:

average annual precipitation;
annual precipitation for individual years (if requested);
average monthly precipitation;
average annual temperature;
average monthly temperatures;
predominant wind direction and its velocity;
average monthly saturation deficit.

If evaporation has been measured at a nearby station by some type of evaporation pan, average evaporation data may also be provided on request. Daily records and, in particular, originals of observers' reports and log books may be consulted in the Central Office. Storage on magnetic tapes and use of automatic station loggers of data facilitate the provision of original data through computer printouts.

3.6 Hydrometeorological instruments, collection and preliminary processing of data

(a) Instruments and data on air temperature

This is mainly measured by a liquid-in-glass thermometer placed in the louvred screen. The routine observation thermometer is filled with mercury and its scale is calibrated in degrees Celsius (°C) from −30 or

$-50°$ C to 40 or 50 °C in markings of 0.5 or 0.2 °C. Temperature may be registered at stations with a maximum accuracy of 0.1 °C.

Some countries use the Fahrenheit scale (°F). One degree on this scale (°F) corresponds to $\dfrac{5}{9}$ °C, and 0 °C corresponds to 32 °F. Thus:

$$°C = \frac{5}{9}(°F - 32),$$

$$°F = \frac{9}{5}(°C + 32).$$

Problem 1

What is the equivalent of 30 °C in °F?

$$\frac{9}{5} \times 30 + 32 = 54 + 32 = \textbf{86}\ °F.$$

Maximum and minimum thermometers (Fig. 3.2) are arranged so as to indicate the lowest and highest temperatures reached between two observations. The most common type of maximum thermometer has its capillary bore constricted below the lowest graduation. When the temperature rises, the mercury is pushed through the constriction and, when the temperature drops, the mercury is prevented from receding by the

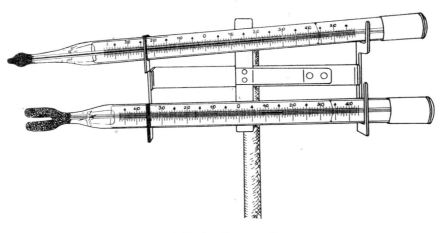

Fig. 3.2. Station thermometers

constriction and remains in its highest position. It is similar to a medical thermometer and, after reading, the mercury is shaken for resetting in the bulb. It is fixed in a nearly horizontal position, with the bulb lower than the other end. The minimum thermometer has a little glass index immersed inside the mercury column or, more often, in the alcohol which replaces the mercury. When the temperature drops, the surface tension meniscus drives the index to its lowest position. When the temperature rises, the liquid flows past the index, which registers the lowest position reached by the upper end of the column. This thermometer should be situated in a horizontal position so that the index does not move by its own weight. After reading, the thermometer is tilted so that the index slips back to the surface tension meniscus.

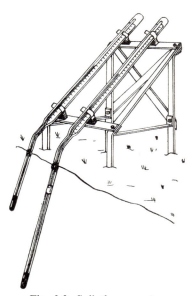

Fig. 3.3. Soil thermometers

(b) Instruments and data on soil temperature

Soil temperature is measured by a thermometer whose mercury bulb is at the end of a long, bent capillary tube which is placed in the ground; the scale remains on the surface (*Fig. 3.3*). It can measure temperatures at depths of 10, 20 and 50 cm, and these are read without removal. For depths of 1 m, a normal straight thermometer is used, attached to a wooden stick and pulled out when read.

Thermometers can also be filled with other substances, such as mercury-thallium. The minimum thermometer may be filled with creosote, toluene, etc. Temperature can also be measured by the change in shape of bimetal strips under the influence of temperature changes. The automatic recording thermometer or *thermograph* is based on this bimetallic principle. There are also *electric* and *gas thermometers*.

(c) Processing of temperature data

Temperatures are read immediately after the screen is opened, the nearest tenth of a degree being estimated first to avoid any changes caused by the observer's breath. The nearest whole degree is read afterwards. The scale reading and the eye of the observer must be in a horizontal line order to avoid any parallax error.

Air temperatures registered at 0700, 1400 and 2100 h are entered in the log book and, later, in the monthly report. This also includes the average daily temperature which is computed as a weighted arithmetic average of all three daily observations, assuming a double weight for the 21-h observation.

Thus:

$$t_{\text{average daily}} = \frac{t_7 + t_{14} + 2t_{21}}{4}.$$

Average daily temperature may also be computed as the arithmetic average of the maximum and minimum temperatures:

$$t_{\text{average daily}} = \frac{t_{\max} + t_{\min}}{2}.$$

This value is, however, departing from the true daily average, for example, as recorded by a thermograph. So, in the USA, the value computed from the $t_{\max}$ and $t_{\min}$ is usually about a degree above the true daily average.

From the average daily temperature may be computed, by arithmetical average, the average monthly temperature and, in the same way, the average annual temperature for a given year; finally, the long-term mean temperature at a given station is ascertained. If the observations cover a specific 30-year period, the long-term mean is called 'normal'.

If the meteorological stations and their average temperatures are plotted on a map, locations with the same air temperatures may be ascertained. Curves defining areas with the same average temperature, mostly for individual months of the year, are called *isotherms* (*see also section 2.2*). Such maps are included in the Climatic Atlas published by the Czechoslovak Hydrometeorological Institute and by many other meteorological services of the world. WMO has published such an atlas for Europe and is now planning a Climatic Atlas of the World.

(d) Instruments and data on air humidity

Air moisture is measured at meteorological stations by a stationary *psychrometer*. This instrument consists of two glass thermometers, one of which has the mercury bulb wrapped in a wick with one end submerged in a container of distilled water. Thus, the psychrometer is composed of one dry thermometer and one wet-bulb thermometer. The temperature of the wet bulb is proportional to the evaporation from the wick which, in its turn, is proportional to the air moisture. From the temperature difference between the two, the air moisture can be calculated, using the *psychrometric tables* (see appendix 6). The instrument can also be arranged so that the air circulates around the thermometers by putting them in a metal case with a small ventilator on a spring or a small electric motor.

A less accurate but more convenient instrument for measuring air moisture is the *hair hygrometer*, which is based on the capacity of the human hair to react to changes in the air moisture by expanding or contracting. The movement is registered by an indicator which shows the degree of relative moisture *e*. The instrument must be calibrated by comparison with the measurements of a psychrometer. The absolute value of air humidity may be determined if the instrument is also equipped with a thermometer. The *hygrograph*, an automatic registering instrument, may be based on this or several other principles. It is usually equipped also with a *thermograph*, thus becoming a *thermohygrograph*, which records the absolute value of air moisture at any moment.

Air moisture can also be determined by measuring the temperature of the dew point. This is done by the so-called *dew gauge* or *condensation hygrometer*.

The difference between maximum and immediate air moisture in millibars or mm Hg, or the so-called saturation deficit, is very often needed in hydrological calculations. Specialized services (meteorological or hydrological) now use data loggers which store the daily values on magnetic tapes. The computing of averages is then a matter of routine computer operation. Monthly or annual averages are most important for hydrologists but it would be too time-consuming for engineers to compute them from daily measurements of air moisture. The average

monthly temperature and air moisture are therefore used, and with
these figures the value of the saturation deficit can be determined from
psychrometric tables. It should be kept in mind that the maximum
water vapour pressure is not a linear function of the temperature, hence
the figure for maximum moisture calculated from averages of temperature must be corrected. The average saturation deficit d will thus be
expressed as

$$d = d + \Delta d,$$

in which d is the value from tables of the saturation deficit, calculated
            from average temperature and moisture,
    Δd = correction.

This correction is given in tables or graphs.

(e) Instruments and data on wind speed

Wind speed, which is necessary for calculating evaporation, for
instance, is most commonly measured by wind vanes (*Fig. 3.4*). The

Table 4

Beaufort scale on Wild wind vane	Wind speed (m/s) at a height of 10 m above ground
0	0—0.2
1	0.3—1.5
2	1.6—3.3
3	3.4—5.4
4	5.5—7.9
5	8.0—10.7
6	10.8—13.8
7	13.9—17.1
8	17.2—20.7
9	20.8—24.4
10	24.5—28.4
11	28.5—32.6
12	> 32.7

Wild wind vane, generally used in Czechoslovakia, is a small tin plate weighing 200 g and measuring 300 × 150 mm, freely suspended on a horizontal axis and fastened to a mast, usually at a height of 10 m, turning in the direction of the wind. The wind vane pivots on its horizontal axis and from its deviation the speed of the wind may be expressed in Beaufort scale degrees. The conversion of wind speed to metres per second from the Beaufort scale is given in *Table 4.*

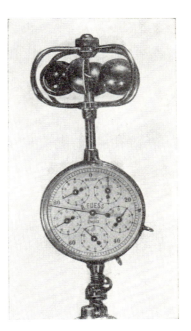

Fig. 3.4. Weather (wind) vane   Fig. 3.5. Anemometer

A more accurate and more common device for wind speed measurement is the *cup anemometer* (*Fig. 3.5*). This instrument is formed by the so-called Robinson cup cross comprising three or four cups fixed to a vertical axis. The rotation of the axis is transmitted to a gauge under the cups, indicating the number of rotations per minute and the speed of the wind. This principle may also be used for the automatically registering *anemograph* which records the direction, speed, and gusts of the wind on a tape. In synoptic stations, the wind speed may also be registered by a Pitot tube instrument, known as Dynes anemometer.

(f) Instruments for measuring evaporation (evaporimeters)

These instruments vary, depending on whether they are to measure evaporation from water surfaces or from the soil and plants (evapotranspiration).

Evaporation from water surface

Evaporimeters in louvered screens

These evaporimeters (or evaporation gauges) may be based on the principle of evaporation from either free water surface or a wetted porous surface. The second type is also known as an *atmometer*. The Bellani atmometer consists of a ceramic disc fixed to the top of a glazed ceramic

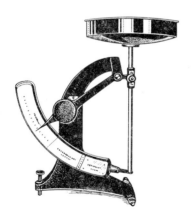

funnel to which water is conducted from a small glass reservoir. The Piché atmometer has, as an evaporating element, a disc of filter paper attached to a graduated glass cylinder (test tube), held upside down and containing water.

Atmometers are attractive because of their simplicity; nevertheless, their measurements are utterly unsatisfactory when compared with evaporation in nature. They are, therefore, seldom used for water resources surveys.

Fig. 3.6. Evaporimeter 'Wild'

In Czechoslovakia, the most common evaporimeters in screens are the Wild evaporimetric scales (*Fig. 3.6*). This is a small pan with an evaporation surface of 250 cm², placed on scales, similar to postal scales. When measuring starts, the pan is filled with 480 cm³ of water which comes up to within 15 mm of the rim. Since the device is located inside a screen, wind speed has no effect on the evaporation and, because of its small evaporation surface, the measurements also differ from actual evaporation in nature. From long term observations by this instrument with a large network of stations, coefficients for conversion of these measurements have been ascertained for calculating evaporation from water surfaces.

Outdoor evaporimeter pans (on dry land or floating)

The use of outdoor evaporation pans with a larger evaporation surface is widespread throughout the world. As in many other countries Czecho-

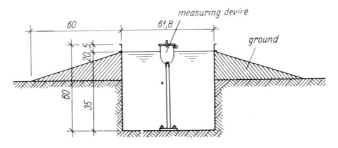

Fig. 3.7a. Pan evaporimeter of the Czechoslovak Hydrometeorological Service

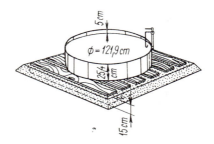

Fig. 3.7b. US Weather Bureau "Class A" pan evaporimeter

slovak hydrometeorological services use pans as network instruments. Until 1963, the pan developed by Roń was used in Czechoslovakia. Its evaporation surface was 2000 cm². The exact level of the surface before and after readings is determined by a siphon and the amount of water which has evaporated is represented by the difference between the water added and precipitation. As this pan evaporimeter has an evaporating surface smaller than an ascertained minimum of

3000 cm², the evaporimeter network in Czechoslovakia now uses pans of 3000 cm² evaporating surface, let into the ground (*see Fig. 3.7a*). The gauge of this evaporimeter is based on a burette, which means that the water level in the instrument is read as a volume and is thus much more precise.

This evaporimeter corresponds in principle with other types of evaporation pans: the USSR sunken pan GGI 3000 with an evaporation surface of 3000 cm², and the U.S. Class A pan with an evaporation surface of 7000 cm². The U.S. evaporation pan, which is situated above the ground (*see Fig. 3.7b*), was also used for international measurement of evaporation during the International Geophysical Year; it is the most widely used instrument in the developing countries.

Fig. 3.7c. Evaporimetric station with a 20 m² evaporimetric basin
(Hlasivo station of the Prague Hydraulic Research Institute)

Fig. 3.7d. Floating evaporimeter

The World Meteorological Organization is at present conducting a world-wide comparison of different evaporation pans, sunken and above the ground, with the aim of introducing an international reference evaporation pan (IREP). Technical Note No. 83 (WMO Publication No. 201. TP. 105) concludes that, while the pans above the ground have several operational and maintenance advantages (such as detection of leakages, easy repainting, etc.), the sunken pans usually give results closer to lake evaporation. However, both types of pan are widely used and both have their advantages and disadvantages (Kohler, 1970).

Evaporation is measured with the greatest accuracy by *basin evaporimeters* with an evaporation surface of between 20 and 100 m². *Fig. 3.7c* shows a photograph of such a basin at the evaporimetric station of the Czechoslovak Hydrological Research Institute at Hlasivo. All types of pan are installed at this station to permit intercomparison and computations of conversion coefficients. Another type of pan is the *floating pan*, which simulates the conditions of natural water bodies. The floating pan of the Czechoslovak Hydrological Research Institute at the above-mentioned experimental station is shown in *Fig. 3.7d.*

Data from pans are transposed into probable lake evaporation by conversion coefficients and by energy-balance computations. Conversion coefficients generally apply only to estimates of *annual* lake evaporation. For the GGI 3000 pan, such coefficients vary from 0.75 to 1.00, and for the Class A pan from 0.6 to 0.8. However, it is recommended that coefficients be ascertained for each climatic region or particular project by local experimental comparison of pan measurements with basin measurements. A complete description and plan of large evaporimetric basins is given in WMO Technical Note No. 83 mentioned above.

Evaporimeters used in measuring evaporation from soil and plants (evapotranspiration)

These instruments are called *lysimeters*. Basically, they are blocks of soil in metal or plastic containers, sunken back into the ground. By weighing them at regular intervals, the water balance of the soil may be established and the evapotranspiration computed. In some lysimeters, the blocks have quite large dimensions ($1 \times 1 \times 1.5$ m, and larger) and cannot be removed from the ground because of their weight; they may then be placed on floaters in concrete tanks containing a liquid and, by means of Archimedes' principle, their weight is constantly registered.

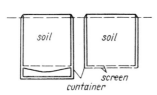

Fig. 3.8. Soil evaporimeter
of Popov

USSR lysimeters

The GGI-500-100 standard soil evaporimeter illustrated in *Fig. 3.8a* consists of an inner cylinder, an external socket-cylinder, a receiver, two handles, and two lifting hooks. The inner cylinder and the evaporimeter socket are made of sheet steel with thicknesses of 2 and 1 mm respectively; the receiver is made of galvanized 0.8 mm steel. The inner cylinder houses a soil monolith. Its interior diameter is 252.3 mm and its height 1 m.

Tags are welded to the rim of the cylinder and hooks are fastened to these when the evaporimeter is lifted and moved.

The cylinder is provided with a cap covering the clearance between the inner and external cylinder walls. The bottom of the inner cylinder, which is removable, has openings to let out the water percolating through

Fig. 3.8a. A set of GGI-500 soil evaporimeters

the monolith, and is fastened to the cylinder by means of three latches. The external cylinder serves as a socket for the inner cylinder. Its diameter and height are 30 and 35 mm larger than the diameter and height of the inner cylinder respectively. The bottom of the external cylinder is compact and waterproof. The receiver is a 30 mm high cylindrical can with an interior diameter of 253 mm. The top is funnel-shape with two openings: one of these, in the centre, collects the water percolating through the bottom openings of the evaporimeter; the other, in the upper part of the funnel near its rim, enables water to be poured into a measuring glass. The inner cylinder, with receiver attached, is sunk into the socket and rests upon the upper rim of the external cylinder by means of the cap.

The GGI-500-50 soil evaporimeter has the same construction, except that the height of the inner cylinder is 500 mm and the height of the external socket-cylinder is 535 mm. *Fig. 3.8b* and *3.8c* show an adaptation of these soil evaporimeters, for weighing them without removal from the soil, as installed by the Czechoslovak Hydrological Research Institute.

The hydraulic soil evaporimeter used at the USSR stations consists of the following components: an inner cylinder, an external case-cylinder, an annular float, a tank, and measuring equipment. The inner cylinder houses a soil monolith. The interior diameter of the inner cylinder is 505 mm and its height is 1500 mm. The cylinder has a removable bottom with openings to release the water percolating through the monolith and

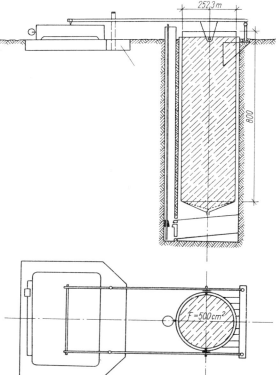

Fig. 3.8b, c. A GGI-500 soil evaporimeter with weighing installation

it is put into an external case-cylinder, resting upon the upper rim of the latter by means of its outer ring. The external diameter of the case is 571 mm; its height is 1305 mm.

There is a receiver at the bottom of the case to collect the water percolating through the holes of the inner cylinder. The case has a ring on its outer surface and this ring rests upon the upper surface of the annular float, which supports the evaporimeter with the monolith in a floating position. The inner diameter of the float is 601 mm; the external diameter is 1620 mm.

The floating evaporimeter system is sunk into a tank installed in the ground and filled with water or a more viscous liquid; the vertical position of the system is regulated by means of ballast loads. Horizontal movement of the system is limited by wire braces.

The measuring equipment of the evaporimeter comprises three micrometers, which fix the vertical position of the floating system, and a piezometer registering the water level in the tank with the float.

In addition, there exists in the USSR a still larger evaporimeter, known as the Great Hydraulic Evaporimeter (BGI).

Data from all types of lysimeters are, for the time being, mainly used in research and cannot generally be adapted for practical hydrological calculations of evapotranspiration from large areas or basins. A method of evapotranspiration computation based on soil moisture measurements and soil water balance is given later in section 4.5f.

(g)   Instruments and processing of precipitation data

An important feature in the installation of the precipitation gauge is its location. In a perfect exposure, the catch of the raingauge would accurately represent the precipitation falling on the surrounding area. This is, however, difficult to attain in practice because of the effect of the wind, and much care has to be given to the choice of site. The effects of wind can be considered under two headings:

(a) the effects on the instrument itself;
(b) the effects of the site on the air trajectories.

In the case of (a), the effect is generally to reduce the amount of water collected. The effects named in (b) are frequently more important and can give rise either to an excess or a deficiency; a reduction in precipitation in one place must result in an increase somewhere else. It should be noted that the disturbance created by an obstacle depends on the ratio of its linear dimensions to the falling speed of precipitation. This effect is reduced, if not entirely overcome, firstly by choosing the site so that the wind speed at the level of the gauge mouth is as small as possible but so that there is not, at the same time, any actual cutting off of rain by surrounding objects, and/or secondly by modifying the surroundings of the gauge so that the airflow across the mouth is made accurately horizontal. It is desirable that all the gauges in any area or country should have comparable exposures and the same criteria should be applied to all.

Wherever possible, the gauge should be exposed with its mouth horizontal over level ground; surrounding objects should not be closer to the gauge than a distance equal to four times their height but, subject to this limitation, a site that is sheltered from the full force of the wind should be chosen, as long as the shelter does not produce larger disturbances in the wind field than the effects which one is trying to avoid. Sites on a slope, or with the ground sloping sharply away in one direction (especially if this direction is the same as that of the prevailing wind), should be avoided. The surrounding ground can be covered with short grass, or be of gravel or shingle, but a hard flat surface such as concrete gives rise to excessive splashing. The mouth of the gauge should be as close to the ground as possible (because wind velocity increases with height) but high enough to prevent splashing. In areas which have little snow and where the surroundings are such that there is no risk, even in heavy rain, of the ground being covered by puddles, a height of 300 mm is widely used. Where these conditions are not satisfied, a standard height of 1 m is recommended.

In very exposed places where natural shelter is not available, it has been found that better results can be obtained for liquid precipitation if the gauge is exposed in the middle of a circular turf wall about 3 m across. The inner surface of the wall should be vertical and the outer surface sloping at an angle of about 15° to the horizontal, with the top level with the mouth of the gauge. Provision should be made for drainage. The main disadvantage of this arrangement is that the space enclosed by the wall is liable to get full of snow in the winter.

An alternative way of modifying the surroundings of the gauge is to fit suitably shaped windshields around the instrument (*see Fig. 3.10*). When properly designed, these enable much more representative results to be obtained than with unshielded gauges fully exposed to the wind. An ideal shield should:

(a) ensure a parallel flow of air over the aperture of the gauge;
(b) avoid any local acceleration of the wind above the aperture;
(c) reduce, as far as possible, the speed of the wind striking the sides, of the receiver;
(d) prevent splashing towards the aperture of the receiver (the height of the gauge mouth above the ground then becomes much less important);
(e) not be subject to 'capping' by snow.

Non-recording raingauges

The basic instrument for measuring precipitation is the daily raingauge. The ordinary daily raingauge usually takes the form of a collector above a funnel leading into a receiver. A receiving area of 1000 cm² is used in some countries but on area of 200 to 500 cm² will probably be found the most convenient. The area of the receiver may with advantage be made to equal 0.1 of the area of the collector. Whatever size is chosen, the graduation of the measuring apparatus must, of course, be consistent with it. The most important requirements of a gauge are as follows.

(a) The rim of the collector should have a sharp edge and should fall away vertically inside and be steeply bevelled outside. The gauge for measuring snow should be so designed that errors due to construction of the aperture by accumulation of wet snow about the rim are small.

(b) The area of the aperture should be known to the nearest 0.5 per cent and the construction should be such that this area remains constant.

(c) The collector should be designed to prevent rain from splashing in and out; this can be done by having the vertical wall deep enough and the slope of the funnel steep enough (at least 45°).

(d) The receiver should have a narrow neck and should be sufficiently protected from radiation to minimize loss of water by evaporation. Weekly and monthly rain-gauges for use at places where daily readings are impracticable should be similar in design to the daily type but with a receiver of larger capacity and stronger construction.

Rainfall gauges in the United Kingdom have an aperture of 127 mm (5 in) diameter, placed so that their rim is 305 mm (1 ft) above ground level. These dimensions were first given in 'British Rainfall' in 1866, when

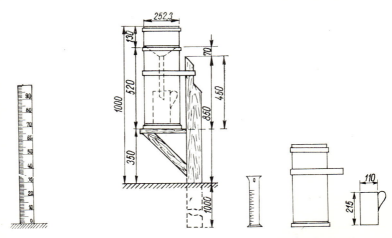

Fig. 3.9a. Snow stake   Fig. 3.9b. Hellman rain gauge

many types and sizes of gauge were in use. The measurement is made according to 'Rules for rainfall observers' (HMSO, London, 1965) at 0900 h. In Czechoslovakia, a gauge of the Hellman type, with a collector of 500 cm^2, is used (*see Fig. 3.9*). The most-used types in other countries, named according to the services which use them, are the US Weather Bureau (collector 203 mm diameter [= 324 cm^2]), and French Association (collector 400 cm^2).

The measurement of the collected precipitation is equally important as gauge catch. Two methods are commonly used for measuring the rain caught in the gauge: a graduated measuring cylinder and a graduated dipstick. A measuring cylinder should be made of clear glass with a low coefficient of expansion and should be clearly marked with the size of gauge with which it is to be used. Its diameter should not be more than about one-third of that of the rim of the gauge, and can with advantage be less.

The graduations should be finely engraved; in general only 0.2 mm should be marked and the line at each whole millimetre should be clearly numbered. It is also desirable that the line corresponding to 0.1 mm should be marked. Where it is not necessary to measure rainfall to this degree of accuracy, every 0.2 mm up to at least 1.0 mm, and every millimetre above that, should be marked; and every 10 mm or every inch line clearly figured. For accurate measuring, the maximum error of the graduations should not exceed ± 0.05 mm at or above the 2 mm graduation mark and ± 0.02 mm below it.

To achieve this accuracy with small amounts of rainfall, it is advisable for the inside of the measuring cylinder to taper at its base. In all measurements, the bottom of the water meniscus should be taken as the defining line and it is important to keep the measure vertical and to avoid parallax errors. It is helpful in this respect if the main graduation lines are repeated on the back of the measure.

Dipsticks should be made of cedar wood, or other suitable material which does not absorb water to any appreciable extent and in which the capillarity effect is small. Wooden rods are unsuitable if oil has been added to the collector to suppress evaporation of the catch; metal or other material from which oil can be readily cleaned must than be used. The rods should be provided with a brass foot to avoid wear and be graduated according to the relative areas of cross-section of the gauge orifice and the receiving can, with allowance for the displacement due to the rod itself; marks for at least every 10 mm should be shown. The maximum error in the graduation of the dipstick should not exceed ± 0.5 mm at any point.

It is also possible to measure the catch by accurate weighing; there are several advantages in this procedure. The total weight of the can and contents should be weighed and the weight of the can then substracted. There is no danger of spilling any and none is left adhering to the can. The common methods are, however, much simpler and cheaper.

The daily raingauge gives the daily amount of precipitation with the highest possible precision. It does not, however, indicate the rate of the rainfall in time, which is best expressed by the intensity of rainfall.

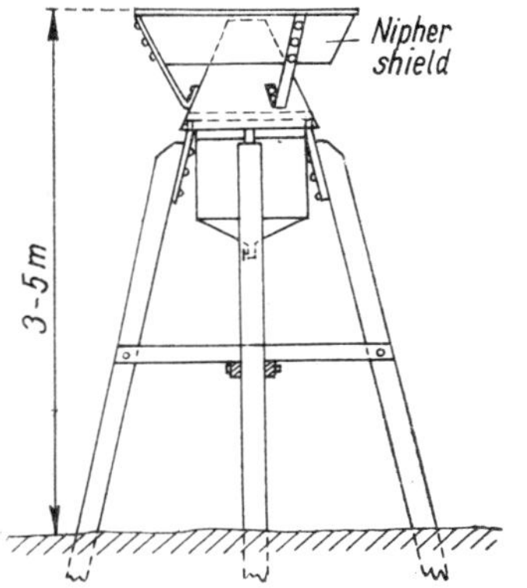

Fig. 3.10. Totalizer

Recording raingauges

In order to measure the intensity of the rainfall, a continuous record is necessary. For this purpose we use recording raingauges, which are of three types: the float, the tilting bucket, and the weighing.

The float type

In this Hellman type of instrument, used in Czechoslovakia and many other countries (see Fig. 3.10), the rain is led into a chamber containing a float; the vertical movement of the float as the level of the water rises is transmitted, by a suitable mechanism, into the movement of the pen on the chart. By adjusting the dimensions of the receiving funnel, float, and float chamber, any desired scale value can be obtained on the chart.

To provide a record over a useful period (at least 24 h is normally required) the float chamber either has to be very large (in which case a compressed scale on the chart is obtained), or some automatic means has to be provided for emptying the float chamber quickly whenever it becomes full, the pen then returning to the bottom of the chart. This is usually done with some sort of siphoning arrangement; the siphoning process should start fully at a definite level with no tendency for the water to dribble over, either at the beginning or at the end of the siphoning. It should not take longer than 15 s. In some instruments, the float chamber assembly is mounted on knife-edges so that the full chamber over-balances. The surge of the water assists in the siphoning process, and when the chamber is empty it returns to its original position. Other rain recorders have a forced siphon which operates in less than 5 seconds, while one type has a small chamber separate from the main chamber to accommodate the rain that falls during siphoning. This chamber empties into the main one when siphoning ceases, ensuring a correct record of total rainfall.

Some sort of heating device should be installed inside the gauge if there is a possibility of freezing during the winter. This will prevent possible damage to the float and float chamber during freezing, and will enable rain to be recorded during that period. A small heating element or an electric lamp is suitable where a supply of electricity is available, but where there is none other sources of power have to be employed. One is a short length of heating strip wound around the collecting chamber and connected to a large-capacity battery. The amount of heat supplied should be kept to the minimum necessary to prevent freezing, because the heat will affect the accuracy of the observations by changing vertical air movements above the gauge and by increasing losses due to evaporation.

The tilting bucket type (see Fig. 3.11)

The principle of this type of recording gauge is very simple. A light metal container is divided vertically into two compartments and is balanced in unstable equilibrium about a horizontal axis; in its normal position, it thus rests against one of two stops, which prevent it tipping over completely. The rain is led from a conventional collecting funnel into the uppermost compartment and after a predetermined amount of rain has fallen, the bucket becomes unstable in its present position and tips over to its other position of rest. The compartments of the container are so shaped that the water can now flow out of the lower one and leave

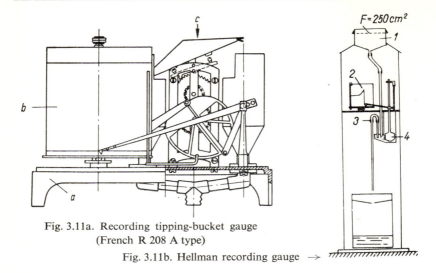

Fig. 3.11a. Recording tipping-bucket gauge
(French R 208 A type)

Fig. 3.11b. Hellman recording gauge →

Fig. 3.11c. Research rain gauge—pit gauge

it empty; meanwhile the rain falls into the upper compartment again. The movement of the bucket as it tips over can be used to operate a relay contact of some kind and produce a record which thus consists of discontinuous steps, the distance between each step representing the time taken for a certain small amount of rain to fall. This amount of rain should be between 0.2 and 1.0 mm if detailed records are required.

The main advantage of this type of instrument is that it can be arranged for recording at a distance or for simultaneous recording of rainfall and river stage on a water-stage recorder. Its disadvantages are:

(a) that the bucket takes a small but finite time to tip over, and during the first half of its motion the rain is being led into the compartment already containing the calculated amount of rainfall. This error is appreciable only in heavy rainfall.

(b) that, with the usual design of bucket, the water surface exposed in it is relatively large, so that evaporation losses can occur, especially in hot regions. This will be most appreciable in light rains.

(c) that, owing to the discontinuous nature of the record, the instrument is not satisfactory for use in light drizzle or very light rain. The time of beginning and ending cannot be accurately determined.

The weighing type

In these instruments, the weight of a receiving can plus the rain which has fallen since the record began is recorded continuously either by means of a spring mechanism or with a system of balance weights. All precipitation is thus recorded as it falls. This type of gauge normally has no provision for emptying itself, but by a system of levers it is possible to make the pen traverse the chart any number of times. These gauges have to be designed to prevent excessive evaporation losses, which may be reduced by the addition of sufficient oil or other evaporation-suppressing material to form a film over the water surface. Difficulties experienced owing to oscillation of the balance in strong winds can be reduced by fitting an oil damping mechanism. The main usefulness of this type of instrument is in recording snow, hail, and mixtures of snow and rain. It does not require the solid precipitation to melt before it can be recorded.

Recording chart

Whether the rainfall recorder operates by the rise of a float, the tipping of a bucket, or some other method, these movements must be converted into a form which can be stored and analysed later. The simplest method of producing a record is to move a time chart by a spring or electrically driven clock past a pen which moves as the float or weighing device moves. There are two main types of charts:

(a) The drum chart. This chart is secured around a drum which should revolve once a day (exactly), once a week, or any other desired period.

(b) The strip chart. This chart is driven on rollers past the pen arm, and by altering the chart speed the recorder can be made to operate for periods ranging from one week to a month or even longer. The time scale on this chart can be large enough for intensity to be calculated with ease.

The movement of a float, bucket, or weighing mechanism can also be converted into an electric signal and transmitted by radio or wire to a distant receiver where records can be made from a number of rain recorders on data-logging equipment.

It is difficult to ascertain which type of gauge is best: the weighing type measures all kinds of precipitation but has a limited time of function; the tilting bucket type is very reliable but gives a discontinuous record; while the float type presents difficulties in measuring the intensity of tropical cyclonic rainfall. Experience is the best guide as to which type to use for local conditions.

Storage gauges (*see Fig. 3.10*)

If access to a gauge is difficult, it is necessary to use a special type called a storage gauge. Storage gauges are used to measure total seasonal precipitation in remote, sparsely inhabited areas. They consist of a collector above a funnel leading into a receiver which is usually large enough to store the seasonal catch. The criteria for exposure and shielding given in previous sections should be taken into account when installing such gauges.

In areas where extremely heavy snowfall occurs, the collector must be placed above maximum expected snow depth. This may be accomplished by mounting the entire gauge on a tower or by mounting the collector on a standpipe which is used to store the catch.

An antifreeze solution is placed in the receiver to convert the snow which falls into the gauge to a liquid state. A mixture of 37.5 per cent of commercial calcium chloride (78 per cent purity) and 62.5 per cent water by weight makes a satisfactory antifreeze solution. Alternatively, an ethylene glycol solution can be used. Though more expensive, the antifreeze solution is less corrosive than calcium chloride and gives protection over a much wider range of dilution caused by ensuing precipitation. The volume of the solution placed in the receiver should not exceed one-third of the total volume of the gauge. A small amount of oil or other evaporation-suppressing material should be placed in the receiver to reduce evaporation.

The seasonal precipitation catch is determined by weighing or measuring the volume of the contents of the receiver. The amount of antifreeze solution placed in the receiver at the beginning of the season must be carefully taken into account with either method.

Measurement of dew

Although the deposition of dew, essentially a nocturnal phenomenon, is not spectacular as a source of moisture, being relatively small in amount and varying locally, it could, nevertheless, be of much interest in arid zones, where it could even be of the same order of magnitude as rainfall.

Since the process by which moisture is deposited on objects largely depends on its source, it is necessary to distinguish between dew formed as a result of downfall transport of atmospheric moisture condensed on cooled surfaces, known as *dewfall* and that formed by water vapour evaporated from the soil and plants and condensed on cooled surfaces, known as *distillation dew*. Both sources generally contribute simultaneously to the observed dew, although at times they operate separately. A further source of moisture is fog or cloud droplets collecting on leaves and twigs and reaching the ground by dripping or by stem flow.

There has been a great tendency to overestimate the average dew over an area, primarily because the physical limits on possible quantities of dew are overlooked. Examination of the energy budget equation reveals that the latent heat of dewfall and/or distillation dew is unlikely to exceed net radiation and should, in fact, be less if sensible and soil heat transfers are taken into consideration. Under favourable conditions, there is a definite limit, at the rate of about 1.1 millimetres per hour, to the average rate of dew formation over an area. However, dew may be substantially increased in local areas where mean temperatures are not horizontally homogeneous and small-scale advection from relatively warmer and moister areas to cooler areas occurs. Moreover, the one-dimensional form of energy flux computation should be modified when applied to isolated plants, since the pattern of radiation and moisture flux is quite different from that of a homogeneous source. This does not mean that the average deposit over a large horizontal area is affected, but only that some parts gain at the expense of others. For several reasons, actual deposition rates will generally be well below the upper limit. In Canada, the highest observed dew on tobacco leaves in southern Ontario was only 1.25 mm, per night, estimated by the blotting technique for a fair number of samples taken during the three tobacco seasons.

Much effort has been devoted, but without much success, to devising means of measuring leaf wetness from artificial surfaces in the hope of yielding results comparable to those for natural conditions. A review of the instrumentation designed for measuring duration of leaf wetness and an assessment of the extent to which various instruments give readings representative of plant surface wetness are given in a document of the third session of the WMO Technical Commission for Agricultural Meteorology.* Any of these devices can only be used as a qualitative guide in any particular situation, or as a crude means of regional comparison, careful interpretation being required in either role. Unless the collecting surface of these gauges is more or less flush with the surface and of very similar properties, it will not correctly indicate the amount of dew the natural surface receives.

Theoretically, the flux technique should give reasonable average values over an area, but lack of knowledge of transfer coefficients under very stable conditions makes it extremely difficult to determine. The only certain method of measuring net dewfall by itself is by a sensitive lysimeter. However, this method does not record distillation dew, since no change in weight accompanies distillation dew. The only generally accepted means of measuring total amount of dew is by the blotting technique, that is, by weighing a number of filter papers both before and after being thoroughly pressed against leaves.

---

* Third session, Technical Commission for Agricultural Meteorology, WMO Document 34, Toronto, 1962.

A brief outline of dew measurement methods developed by Duvdevani (1947) and Kyriazopoulos is given in the WMO Guide to Meteorological Instrument and Observing Practices.*

Telemetering and recording for machine processing

There exist several instruments for telemetering the precipitation observations and registering them in a manner suitable for machine processing. They are described to a certain extent in the WMO Technical Notes No. 74 (Data-processing by machine methods, WMO-No. 189. TP. 95) and No. 82 (Automatic weather stations, WMO-No. 200. TP. 104).

The most usual operational instrument in the USA is a Fisher & Porter punchtape recorder combined with a weighing registering precipitation gauge, which has a 16-channel recording tape and registers accumulated rainfall in time intervals of 5, 6, 15, 30, or 60 min, or 6 or 12 h.

A more detailed description of modern systems for collection, transmission, and processing of data, of automatic weather stations and of other sensors is given in section 5.4.

The radar measurement of rainfall

The fact that radar is capable of detecting and locating precipitation continuously makes it valuable for many hydrological purposes. Its application is becoming an essential adjunct to precipitation gauge networks. It should be borne in mind, however, that radar cannot replace a basic raingauge network, for two main reasons: the data presented by radar display are basically an instantaneous areal sample and it is not easy to relate the brightness of the echo on the radarscope to rainfall intensity. Automatic, semi-automatic and manual systems are used today throughout the world, but the aim is to obtain digitized output for computer processing. Nevertheless, before any investment in radar equipment for such purposes is made, its limitations must be fully understood.

The hydrological range of radar is defined as the maximum range within which the relationship between the radar-echo intensity and the

---

* Guide to Meteorological Instrument and Observing Practices, WMO No. 8. TP. 3, 2nd ed., 1961, chapter VII.

rainfall intensity as measured by precipitation gauges remains reasonably valid. This relationship has been shown to be good up to 60 nautical miles, with decreasing accuracy to 100 nautical miles. Usually, a long-wavelength radar (100 mm) is employed because a shorter wavelength is significantly attenuated in passing through heavy precipitation. A detailed description of the problems of this method is outside the scope of this book. The reader is directed to WMO publications (WMO/IHD Project Reports No. 5, 'Radar Measurement of Precipitation for Hydrological Purposes' by E. Kessler and K. E. Wilk, and No. 9 — 'Hydrological Requirements for Weather Radar Data' by A. F. Flanders).

Errors of observation of precipitation and their correction

Errors in the measurement of precipitation depend on two things:

(a) the accuracy with which a gauge measures the amount of precipitation which would have fallen per unit area of the ground at the gauge site had the gauge not been there;

(b) how well the catches of a number of gauges over an area can represent the total volume of water precipitation on the area.

In general, precipitation gauges underestimate the precipitation that would have fallen on the ground in the absence of the gauges. The magnitude of the underestimate depends on the type of equipment and the observing procedures used, the height above ground-level of the receiving aperture of the equipment, the exposure characteristics of the site, the type and intensity of the precipitation, and on other concomitant meteorological elements, notably wind.

Research carried out in the USSR has shown that, depending on the relevant local conditions and the degree to which gauges are shielded from the wind, the catch efficiency of gauges varies widely (on average over a period of years) from place to place. Observations taken at several hundred stations situated in various parts of the country suggest that the correction which needs to be made to annual total precipitation as measured by raingauges varies between 17 and 56 per cent (usually between 20 and 30 per cent). These corrections apply to raingauges with an area of 500 cm² and a receiving aperture situated 2 m above ground-level, on sites partially protected from the wind. The corrections necessary for gauges sited in places protected from the wind are close to the lower limits and those for instruments sited in the open are close to the upper limits.

Special experimental programmes are needed in many parts of the world to evaluate the catch efficiency of precipitation gauges under the specific local climatological and other relevant conditions. *Fig. 3.10c* shows a special Czechoslovak experimental gauge, located at ground-level in a grid — the so-called 'pit gauge'. Research with such instruments is conducted in the UK, USSR, Czechoslovakia and other countries.

For storage precipitation gauges, errors of observation depend, in addition to the above sources (called, for short, *catch efficiency*), on evaporation, freezing, orifices not being properly orientated, narrowing of the orifice by rime or snow deposit, capping of the orifice by snow, complete covering of the apparatus by snow, or leaking of the container.

Provided reasonable care is taken in the readings, the errors involved in measuring the catch once it has been collected in the gauge are small compared with the uncertainty due to the effect of the exposure of the instrument. Daily gauges should be read to the nearest 0.2 mm and preferably to the nearest 0.1 mm; weekly or monthly gauges should be read to the nearest 1 mm. The main likely sources of error are the use of inaccurate measures or dipsticks, the spilling of some of the water when transferring it to the measure, and the inability to transfer all the water from the receiver to the measure.

In addition to these errors, losses can occur by evaporation. These are likely to be serious only in hot dry climates and for gauges which are visited only at infrequent intervals. They can be reduced by placing oil in the receiver (this forms a film over the surface of the water) or by designing the gauge so that: (a) only a small water surface is exposed, (b) the ventilation is small, and (c) the internal temperature of the gauge is not allowed to become excessive. It is also necessary to ensure that the receiving surface of the gauge is smooth, so that the raindrops do not adhere to it. It should never be painted.

In winter, when rains are often followed immediately by freezing weather, damage to the receiver, and the consequent loss by leakage, can be prevented by the addition of an antifreeze solution. This again mainly applies to gauges visited infrequently. Allowance for the solution added must of course be made when measuring the results. All gauges should be tested regularly for possible leaks.

On the other hand, a *float type recorder* can have the following causes of errors.

(a) During very intense rainfall (monsoon), the float chamber's frequency of emptying can be so high that the emptyings overlap and an error may occur. This happened in the evaluation of rainfall from the hurricane *Flora* in the Carribean region carried out by the author. Fortunately, this kind of error can easily be discovered by a comparison with the regular raingauge reading, which was what happened in the case mentioned above.

(b) The clock mechanism of the gauge may not mark the right time. This error can easily be discovered and corrected by the observer, who has to make time marks by hand on the registration paper according to an accurate timepiece.

(c) The float chamber may not be watertight. In this case, the line marked by the pen will not be horizontal (or vertical) during a period without rain.

(d) The axis of the registering drum and that of the float may not be parallel. A simple formula:

$$i = \frac{i'}{1 - i' \tan \alpha}$$

where i = real intensity,
    i' = measured intensity,
    α = angle of non-parallelism,

may be used for corrections.

(e) The axis of the registering drum may not be parallel with the surface of the drum. This error can be ascertained by making a horizontal mark every hour by the pen's maximum and minimum positions. After a complete revolution, these marks should make two parallel equidistant straight lines. If the lines are not parallel or are sinusoidal curves, the instrument is unfit for registration and must be discarded.

(f) Finally, if the registration chart has not been placed properly on the drum (i.e., with the horizontal lines of the chart not perpendicular to the axis of the drum), the correction of the observation can be achieved by the procedure described in (d).

Instructions to observers

In order to minimize the errors resulting from the human element and instrument malfunction, clearly written instructions must be provided to all observers. These should contain guidance and direction on the following matters.

(a) A brief description of the instruments, with diagrams.

(b) Routine care and maintenance of instruments, and the action to be taken in the event of serious breakage or malfunction.

(c) The procedure for taking observations.

(b) The times of routine observations.

(c) Criteria for beginning, ending, and the frequency of special non-routine observations (for example, river-stage observations while the water-level is above a predetermined height).

(f) The procedure for making time checks and putting check observations on charts at stations with recording instruments.

(g) Filling in the field notebook or station journal.

(h) Filling in the report forms, including methods of calculating means and totals with appropriate examples.

(ı) Sending report forms to the central office.

Such written instructions should be supplemented with verbal instruction to the observer by the inspector at the time the instruments are installed and at regular intervals thereafter.

Snow pack measurement and interpretation

The measurement of snow for hydrological purposes consists of measuring the depth of freshly fallen snow (snowfall) and the snow cover on the ground and their equivalent in water depth.

There are several manuals describing these practices; among the international ones are the WMO Guide on Instruments and Methods of Observation, the WMO Guide to Hydrometeorological Practices, and a newly prepared manual of the International Commission on Snow and Ice of the IASH. We are, therefore, only summarizing them on the basis of the last-mentioned manual.

All measurements related to snow accumulation and ablation depend greatly on the characteristics of the site, and in particular on the exposure to wind and sun and on the frequency and velocity of the wind. Unless there is some particular motive in choosing a special exposure, a site for measuring snow should be:

(a) horizontal;

(b) open to snowfall and insolation;

(c) sheltered against strong winds and drifting snow;

(d) distant from any objects which could cause an excessive snow deposit.

These conditions partly contradict each other, and in most cases a compromise has to be found. An optimum site is a flat area of the order of 50 to 100 m in diameter and surrounded by sheltering objects (such as trees or buildings) no higher than about one-eighth of the diameter.

Snowfall, i.e., solid precipitation, may be measured either by catching falling snow on its way to the ground with a receiver or by measuring the amount of fresh snow deposit on open ground.

Instruments of the receiver type serve the purpose of measuring snowfall and rain as momentary values (recording types), as hourly or daily values (storing types), or as monthly or seasonal values (storing types). As the measurement of snow is more affected by deficiencies than that of rain, the instrument should be primarily built for snow.

A fresh snow deposit on open ground is subject to the natural influences on a site, which are evaporation and condensation, loss by melting, and accumulation and erosion by wind. A daily measurement therefore represents the daily snow accumulation rather than the pure precipitation. Besides the fact that the accumulation is often of greater interest than precipitation (in the hydrological sense), the measurement of the deposition on open ground is in many cases more reliable than the catch with precipitation gauges. This is valid in particular for areas with a proper snowfall season and a lasting snow cover.

The measurements are made either by the snow-board method or by sounding on a cleared plot. A board of at least 300×300 mm is set flush with the snow surface in a measuring site and marked with a thin stick at the beginning of the observation day. After 24 h, the *fresh snow depth* is sounded with a calibrated rule down to the board. To take the *water equivalent*, a cake of snow is cut out vertically with a metal cylinder of known cross-section (usually 500 cm^2) and weighed. Measuring the accumulation on a daily cleared plot is an emergency procedure.

Whereas measurements of snowfall and fresh snow deposit are of primary interest with respect to the precipitation characteristics of an area, measurements of snow-cover data deal with the area snow coverage and its development from the beginning of a seasonal snow cover to its very end. Gains by drift and losses by erosion and run-off are included in the budget.

The total snow depth is measured by snow stakes or by snow courses. On a suitable site, a calibrated stake is mounted in a position which can easily be inspected without a close approach. Any trespassing within a radius of about 5 m should be prevented by a loose wire fence. The calibration should be in metres, decimetres, half decimetres and centimetres, easily distinguishable from the fence. The readings are taken by sighting over the undisturbed snow surface to the stake.

Snow courses

If the true average snow depth over a large area with locally variable snow conditions is to be obtained, a great many observations must be taken. The area in question may be a watershed or a part of it, as, for example, a specific exposure or zone of altitude.

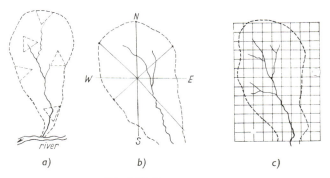

Fig. 3.12. Snow courses

A network of observation sites — a so-called *snow course* — is established. There are three kinds of arrangement (*see Fig. 3.12*).

(a) A number of sites (five, for example) are chosen, each of which is representative for certain dominant conditions. On each site, several measuring points (say, for example, ten) are selected in triangular paths.

(b) A large number of single measuring points (say, perhaps, 50) are installed along paths which can be easily followed. The distance between the measuring points is of the order of 50 to 100 m. All characteristics of the area should be represented in a well balanced manner. These are called linear snow courses and they may be disposed in the main wind-rose directions.

(c) The single measuring points are disposed along paths in a dense rectangular grid.

Snow courses are mostly utilized in connection with *snow survey*, which includes the measuring of the water equivalent of the snow pack.

In inaccessible areas such as dangerous slopes or remote places, snow depth measurements are often taken with field glasses or telescopes from stakes provided with short crossbars every 500 mm. If the inspection is made from an aircraft, a photographic record is to be recommended.

Water equivalent of the snow cover

The water equivalent of the snow pack is the key figure for the hydrological situation of a snow-covered watershed. It can be ascertained by sampling in open pits dug in the snow or by drilling from the surface. Snow-pit studies are worth while, above all, if other snow characteristics like stratification, snow temperature, local density values, etc., are included in the observation.

Core drilling from the surface

This method is widely applied on a routine base for the snow survey in various countries. The equipment for core drilling consists of:

(a) a sampling tube with a toothed cutting head (steel) for clockwise operation;
(b) extension tubes (steel, duraluminium, or plastic);
(c) a handle (turning lever)
(d) weighing scales together with a sample carrier;
(e) a sounding rod;
(f) cleaning equipment, tools, etc.

The inner diameter of the cutting head, i.e., the diameter of the core, ranges between 37.7 and 90 mm. For the tubes, an inner diameter slightly above that of the core is selected. They are calibrated for measuring snow depth and often provided with slots over the whole length for inspecting the length of the core. The sampling procedure is as follows.

(a) A sampling point within the area is selected and prepared the previous summer. It must be levelled, cleared, coated with soft soil, etc.

(b) The position in relation to a reference point (marker) is noted on a sketch.

(c) A previous sounding is made with the sounding rod in the vicinity of the sampling point in order to check snow depth and quality (ice layers).

(d) The sampler is pushed down vertically in a continuous movement, turning it clockwise. Push and twist are adjusted to the snow quality, extensions being added if necessary.

(e) When the ground is reached (it will be recognized from the previous sounding), a plug of soil is picked up.

(f) The outside snow depth is read on the tube.

(g) The sampler is withdrawn and inspected at the bottom. The length of the core is checked, the soil plug (if any) removed and measured.

(h) The full sampler is weighed (the weight of the tubes when empty must therefore be known).

(i) The sampler is emptied and prepared for the next sounding (inspect the cutting head!). Snow cores and mud plugs are deposited outside the observation area.

All readings and observations are to be noted immediately. On a snow course, water equivalent measurements and snow depth measurements are often combined. The mean density is thereby assumed to be a property scattering less than snow depth.

A direct measurement of the water equivalent in measuring the pressure exerted on the ground by means of electrical pressure plates or hydraulic pressure cushions (pillows) has been successfully attempted.

Type of station	Type of observation	Frequency and date
Manned stations	fresh snow depth fresh snow water equivalent total snow depth extent of coverage water equivalent of snow cover	daily standard hour morning observation (between 0700 and 0900) fifteenth and last day of month

Snow courses, unmanned stations	snow depth
	water equivalent of snow cover
	storage gauges
		end of month
		end of month, at least twice
		   a year (31 March,
		   30 September)

Area inspections	observation of snow lines extent of snow coverage	31 March 30 June 30 September 31 December

Gamma rays emitted by a radioactive source such as, for example, cobalt 60, are absorbed by water or ice according to the water equivalent of a snow layer placed between the source and a receiver. For obvious reasons the application of this method is restricted to a few important representative sites. Nevertheless, in Czechoslovakia this method has been developed and is used at the operational level. Natural radioactivity of soil presents another promising approach to areal and automatic measurement of snow water equivalent. This method has been successfully developed in the USSR.

The area of the snow cover can be estimated from the ground or observed from aircraft and satellites.

Standard times and dates for snow measurements can most easily be presented in tabular form, as on the previous page (see also ICSI/WMO/UNESCO Manual, 1969).

3.7   Hydrometry, measurements of streamflow

Under the term hydrometry is generally understood a network of stations measuring the following elements: river and reservoir stages, discharges of rivers (streamflow), sediment transport, chemical quality and temperature of water, and characteristics of ice covers on rivers and lakes.

The WMO Technical Regulations, Volume III, Operational Hydrology (WMO Publ. No. 49, 3rd. ed., 1971) recognize as hydrological observing stations or hydrological stations for specific purposes: principal and secondary hydrometric stations and groundwater stations. A climatological station established for hydrological purposes is also considered as being a hydrological station.

The design of a hydrometric network depends on many complex conditions of which the economic constraint is certainly not the least. Nevertheless, in particular for the stream-gauging network, some principles of design have been ascertained. These may be summarized in accordance with the WMO Guide to Hydrometeorological Practices and with respect to the minimum density of such a network as follows.

I.   Flat regions of temperate, mediterranean, and tropical zones

4−10 stations for 10 000 km^2; 1 station for 1000−2500 km^2.

But for such countries where it is difficult to achieve, for the time being, the required density because of sparse population, lack of development of communications facilities, or for other economic reasons, the density of the stream-gauging network may be reduced to:

1 station for 3000 $-$ 10 000 km^2.

II.   Mountainous regions of temperate, mediterranean, and tropical zones

In mountainous regions, it is most desirable to have stations distributed in altitude zones of approximately 500 m per zone with minimum density of:

10 $-$ 30 stations for 10 000 km^2; 1 station for 300 $-$ 1000 km^2.

For countries where it does not seem possible to achieve, for the time being, the required density because of the small density of the population, the lack of development of communications facilities, or for other economic reasons, the density of the stream-gauging network may be reduced to:

1 station for 1000 $-$ 5000 km^2.*

On the other hand, for small mountainous islands less than 20 000 km^2 with very irregular regimen and very dense stream network, the minimum density is:

1 station for 140 $-$ 300 km^2.

III.   Arid and polar zones

0.5 $-$ 2 stations for 10 000 km^2; 1 station for 5000 $-$ 20 000 km^2, depending on the feasibility.

Such norms are not applicable to great deserts with no defined stream networks (such as the Sahara, Gobi, Arabian, and Korakorum deserts) and great ice fields (Antarctic, Greenland, Arctic islands).

Wherever the density is less than 1 station for 4000 km^2 for categories I and II, a region of the order of 3000 km^2 should be instrumented to the more stringent standards of these categories in order to obtain information on the variability of runoff. This region should coincide with the 3000 km^2 region mentioned under '*Precipitation Stations*' (WMO Guide to Hydrometeorological Practices).

---

* Under very difficult conditions this may be extended to some 10 000 km^2.

As regards the preceding paragraph, the stations should be equally divided into two categories: large river stations and stations on small streams, except for certain countries which will have only small rivers.

The value of A, the area of catchment which divides a main stream from the small-stream network, is defined as follows:

for regions of category I, $A =$ 3000 to 5000 km^2;
for regions of category II, $A =$ 1000 km^2;
for regions of category III, $A =$ 10 000 km^2.

In general, stations should be sited on all streams where the catchment area reaches or is greater than A. In effect, the minimum network permits nearly complete coverage of all large streams as defined above, unless the stream pattern is highly irregular. However, the placing of an excessive number of gauging stations on large streams only should be avoided.

In applying this rule, it may be noted that, like the raingauge network, the stream-gauging network will meet most immediate needs only if certain principles of installation and use are followed.

Lake and reservoir stages

At these stations, observations should be made on stage, temperature, surge, salinity, ice formation, etc.

Stations should be installed on lakes and reservoirs with surface area greater than 100 km^2. As in the case of rivers, the network should sample smaller lakes and reservoirs as well.

(a)   Discharge and stage measurements

Almost every hydrological service has its own instructions on discharge measurements and operation of stream-gauging stations. There is an international recommendation for standards in this field; the International Organization for Standardization (ISO) published the following recommendations or drafts:

Draft Recommendation 1071, Establishment and operation of a gauging station and determination of the stage-discharge relation;

ISO/R. 748 – 1968, Measurement of liquid flow in open channels by velocity area methods;

Draft ISO/DR. 1438, Measurement of liquid flow in open channels by the use of weirs and flumes;

ISO/R. 555 — 1966, Liquid flow measurement in open channels, dilution methods for measurement of steady flow, Part I — constant rate injection.

The work of ISO is progressing in almost every respect of importance in this field.

National standards, which are often the origin of ISO recommendations, are also available in many countries. In the UK, the British Standard Specification 3680, 'Measurement of liquid flow in open channels,' covers the following subjects:

broad crested weirs;
weirs and flumes;
dilution methods;
velocity area methods.

The Czechoslovak Hydrometeorological Service also provides instruction corresponding to international standards in manual.

The above publications may serve as exhaustive reference; they are, however, mainly directed to the hydrological services.

The following descriptions of stream-gauging methods and related measurements, while in no way departing from the principles included in the international standards, are directed rather to engineers concerned with hydrometry in connection with design or construction of water resources or other projects on streams. They largely reflect the Czechoslovak and the author's personal experience.

For engineering projects, especially in construction along streams, the stream-gauging cross-section is usually determined purely by structural and economical considerations. However, in conducting general hydrological surveys on a stream or a reach of it, important for a given purpose, it is necessary to choose a suitable cross-section for stream-gauging according to certain principles. In choosing this site, the same premises as used for hydrometeorological station selection, namely the representativeness of the station and density of the network, are not sufficient. The suitability of the site must be considered with respect to its location on the stream (it should be on a straight reach, beneath the mouth of a tributary, beyond the backwater of a reservoir, etc.), and also from the standpoint of the watershed, the runoff of which the cross-section represents.

In selecting a suitable site, the stream is considered according to reaches characteristic of the morphology of the stream bed and valley aspects, and also from the runoff regime aspect, especially in periods of flood.

The beginning or end of these characteristic reaches is usually situated:

(a) below the confluence of an important tributary. It can be a brook in the case of a smaller stream.

(b) where a narrow and confined stream bed enters a wider valley and the water floods the floodplain.

(c) where the above-mentioned wide valley ends and the stream flows into a narrow, confined valley.

(d) where there is an obviously sharp change in the slope and the water flows either more rapidly or more slowly.

(e) in broader valleys where there is a basic change in the crops grown along the stream.

(f) above and below settled areas. This mainly concerns smaller communities; streams flowing through larger towns are usually fully regulated.

(g) at sites of water management and transport structures—above inlets into reservoirs, above bridges, above weirs but outside backwater, above intakes of canals, etc.

(h) in places suitable for possible future water management and transport projects.

After the characteristic reaches have been defined, the stream-gauging cross-section is established at the beginning, in the middle, and at the end of each reach.

The sites having been chosen, the basin corresponding to each cross-section should be investigated.

Depending on the shape of the catchment, the site will either remain in the place originally selected or be moved so that the shape of the catchment is homogeneous and different sub-basins are properly interconnected. Thus, for example, if one sub-basin runs into the basin beneath it in a narrow tongue (*Fig. 3.13*), it is better to shift the gauging cross-section so that, if possible, this tongue is eliminated. This helps to simplify many hydrological calculations.

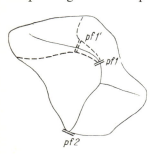

Fig. 3.13. Gauging sites

A stream-gauging station measures either the stage and discharge or only the stage. In the former case, a stage-discharge relation is developed and periodically checked.

(b) Measuring river stages

River stage is the elevation of water surface of the stream with respect to a datum. Wherever possible, a stage-measurement cross-section is located on a straight reach of stream. It is advantageous if all discharges can be measured in its immediate vicinity (even floods).

The bottom and banks of the channel at stage gauging stations should be stable; if not, an artificial control must be built. If the station is on a tributary, it must not be affected by the backwater of the main stream, especially during floods.

Bridges with a large number of piers or bridges which cross the stream at a considerable angle are not suitable as a site for stage measurement.

Otherwise, bridges are often used for this purpose, because it is safe to attach the gauge to a pier, it is easily accessible and discharge may be measured from the bridge even in time of flood.

Instruments

The basic instrument of stage measuring is the *staff gauge*, a measuring scale which takes in the lowest and highest probable water level in the cross-section. It is graduated in 1 or 2 cm sections, in the latter case the odd numbers being estimated (*Fig. 3.14*). Zero on the staff gauge is to be situated at the lowest point of the scale so that all readings are positive. Some old staff gauges in Czechoslovakia, and many in other countries of central Europe, have zero at the so-called 'normal' water level, so that they give a negative reading when the stage is lower than 'normal'. Such an arrangement of zero datum is not recommended and is now, in fact, abandoned, since the 'normal' stage varies with the evolution of the stream bed. In addition, the transmission of negative and positive readings through telecommunication channels may often cause errors which, particularly during an emergency, may have disastrous consequences.

Zero on the scale must be levelled and in smaller streams the level is secured by one fixed datum point beyond the reach of the water. On larger rivers, in Czechoslovakia, three such fixed datum points, connected to the state levelling, are required.

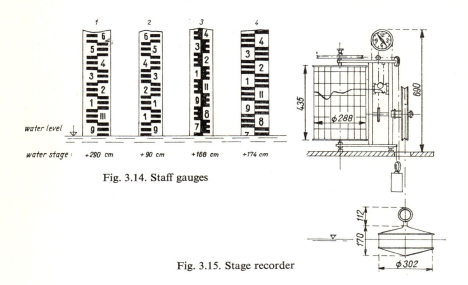

Fig. 3.14. Staff gauges

Fig. 3.15. Stage recorder

If it is not possible for one staff gauge to cover all water stages, gauging is done in several sections at different locations and elevations within the cross-section.

The scale zero for all sections must be the same as far as elevation is concerned.

Staff gauges may be either vertical or inclined. It is preferable that they are not exposed to the swift current so that they are not damaged by floating objects, particularly ice.

The staff gauge is either wooden (hardwood, 16 to 20 cm wide, 50 cm thick, usually with the scale carved out) or metallic, often enamelled. In Czechoslovakia, full metres are marked in red with roman numerals, decimetres in black with arabic numerals.

According to Czechoslovak stage gauging instruction, regular observations are made at 0700, 1200 and 1800 hours, local time. In winter, observations may be shifted to 0800, 1200 and 1700 hours. Special readings are taken more frequently during sharp changes in the water level (during floods), sometimes as often as every ten minutes, in order to record the highest (peak) water stage in hours and minutes. Time is indicated in local time. If the peak occurs at night, without being observed, the mark of the peak must be sought in the vicinity of the water gauge in the morning. It can be found mainly from traces of sediment. Such traces may be easily found on trees. It is recommended that such high-water marks can be recorded permanently by paint on houses, piers, etc.

Fig. 3.15a. Stage recorder on pipe (courtesy of Électricité de France – Grenoble)

In the USSR, a special crest stage indicator is widely used, connected to a staff gauge. A float is used to indicate the maximum level reached. In view of the large number of recorders, this indicator is not generally used in Czechoslovakia. Nevertheless, as experience in the United States also shows, such crest stage indicators are very useful for indications of stage at isolated points, far from hydrometric stations.

A continuous chart of stage record is given by an automatic recording instrument (limnigraph). Types manufactured in Czechoslovakia are based on the principle of a float (*Fig. 3.15*). The float follows the movements of the water surface and the movement is transferred to a gear which reduces the range of motion (the range of fluctuation of the water

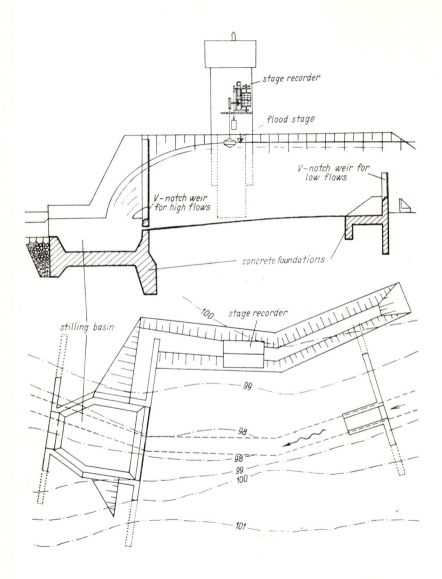

Fig. 3.16. Stream-gauging station on small streams

level is generally much greater than the height of the registration paper); a pen records on a chart fixed around a drum turned by clockwork once a week (or once in 24 h). In large stations, the instrument is located in a shelter on the bank and the float is located in a stilling well connected to the stream by a pipe (*Fig. 3.15a*). The intake of the pipe is protected against floating objects, and a filter (several perforated partitions or gravel) is placed in the pipe in order to suppress the effect of dynamic waves.

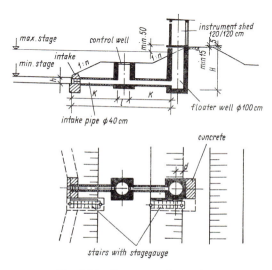

Fig. 3.16a. Stage-recorder station

This layout is used in Czechoslovakia for large IBA-G and IBA-G (weight-driven clock) recording gauges, which are similar to the German OTT XX. Simpler recording gauges in Czechoslovakia (the IBA Pi and Javorský—Frič types) are used on smaller streams and at temporary stations. They may be mounted on a steel pipe with a minimum diameter of 200 mm and fastened on it by screws. The pipe may be attached to a bridge pier or clamped against the bank. Gravel fill at the bottom of the pipe helps to suppress dynamic surface waves.

Simple recording gauges may also be portable and can be very easily installed, even on a provisional wooden weir. They are therefore often used for temporary engineering measurements and for research purposes.

A hydrometric station on a research basin in Czechoslovakia (*Fig. 3.16*) shows the layout of the stage recorder in relation to the measuring weir.

The recording chart for the above stage recorders usually has thick horizontal lines correspoding to 50 cm of level variation (scale 1 : 10); thinner lines divide this into 10 cm sections and the thinnest lines indicate every 1 cm. Vertical lines represent one hour. The chart covers 7 days and 18 hours, but it can be used for a whole month by vertical adjustment of the pen, once the drum has made one revolution. Thus four weeks may be recorded, one under the other, on a single chart. This procedure is, however, used only in emergency, which may occur quite often in developing countries.

When a new chart is fastened in the instrument, the pen must be set exactly in time and stage co-ordinates. Time setting is possible by reverse

Fig. 3.16b. Bazin gauging weir

drum movement. A short line on the chart, marked by moving the float wheel manually, indicates the exact time of setting of the instrument.

The pen must also be set on the exact water level or a datum mark must be indicated on the chart. Thus, every stage recorder is set according to a staff gauge, which provides the most precise basic data about the water stage, against which the recorder must be checked.

In order to prevent possible malfunctioning of a recorder during an extreme flood, caused by the insufficient size of the recording chart, most types of stage recorders can automatically reverse the direction of the pen when it reaches the end of the chart. Such an arrangement is particularly easy with a horizontal axle of the recording drum. The US recorder Leupold—Stevens, the German OTT X, the Russian Valdayi, and the Czechoslovak METRA UL-501 have this arrangement.

During the winter, owing to icing, recorders are often disconnected (unless heated) and stages are measured only by staff gauges.

Several other principles for measuring stream stage may be used. In particular, change of air pressure in a pipe, caused by different hydrostatic heads at different stages, has been and is still used. The so-called 'bubble-gauge' is widely used throughout the world. Such instruments are produced commercially in France (Neyrpic, Grenoble) and in the USA (Leupold Stevens). The advantage of this instrument is that it does not necessitate a stilling well or any other structure in the stream bed. This stage recorder is also very suitable for advanced digital systems of recording, such as punched paper-tape or magnetic tape which, in its turn, facilitates telemetering and automatic long-distance telecommunication. Nevertheless, the classic float-type instruments are simpler and also permit telecommunication as well as digital recording. Such an instrument was constructed in France under the commercial name of 'Limniphone'; this gives recorded stage information on interrogation through normal telephone networks (*see also section 5.4*).

(c)   Discharge measurements

Hydrometric stations of the Czechoslovak Hydrometeorological Service are divided into three classes. Class I stations are all equipped with recorders and have a stable cross-section in which the whole discharge is concentrated. Class II stations are on larger rivers but may have the channel bifurcated and the cross-section may include dikes and levees; they need not always have automatic recording instruments. All other stations, most of which are on relatively smaller streams, are in class III.

Selection of a cross-section for measuring discharge

If the hydrometric station is to serve for discharge measurement, it must fulfil additional conditions. In swift, deep streams such sites are selected where the stream is broad and not too deep; in streams with slow current, narrow reaches with deep and swift flow are selected. The cross-section must also be outside any natural or artificial backwater. For developing the discharge rating curve of the cross-section, it is often necessary to measure the slope of the water surface; hence, the selected site should permit such measurement on both banks without any difficulty.

Methods of measuring discharge

Discharge is measured directly or indirectly. *Directly* it may be measured volumetrically (by a container). With the help of hydraulic structures, such as weirs and flumes, and using their volumetric calibration, discharge may also be measured directly.

Discharge is measured *indirectly* by the area-velocity method with:

(a) direct measurement of average velocity v_a by a float, a current meter, Pitot tube, or other velocity-measuring instruments;

(b) calculation of the average velocity by a formula such as the Chézy (Manning) equation.

In the area-velocity method, the discharge Q is computed by the equation

$$Q = Av_a$$

where A = the area of the discharge profile, and

$\quad v_a$ = average cross-sectional velocity.

On smaller streams, discharge is usually measured directly by setting up an artificial control in the form of a *weir* or a *flume,* or by a combination of these methods. A Thomson triangular weir is most advantageous

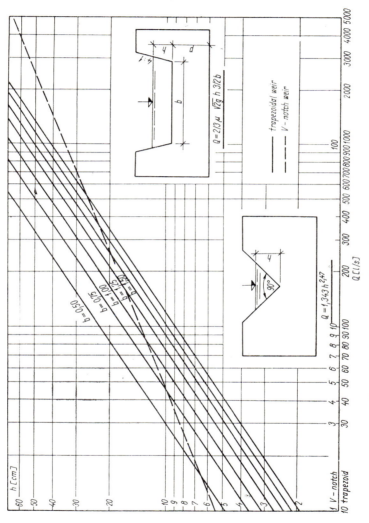

Fig. 3.17. Nomograms for V-notch and trapezoidal weirs

(*Fig. 3.17*); its discharge rating curve is a straight line in logarithmic co-ordinates. The weir is usually made of steel plate set into wooden or cement supports. The notch is in the form of a triangle. The crest is sharp and thin-plated. (The discharge rating curve of the weir in *Fig. 3.17* was ascertained at the Hydraulic Research Institute in Prague.) This weir

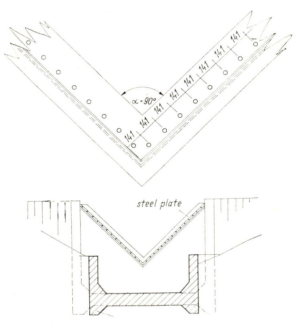

Fig. 3.17a. V-notch of Fig. 3.16 (detail)

measures discharges up to 135 l/s with a head of 400 mm. For larger discharges, it is better to use the rectangular Bazin weir (*Fig. 3.16b*) or the trapezium weir (*Fig. 3.17*). Such weirs measure discharges up to 3 m³/s.

Details of the construction of the Thomson weir are shown in *Fig. 3.17a*. The hydrometric station at the experimental basin of the Slovak Academy of Science's Institute of Hydrology and Hydraulics is shown in *Fig. 3.18*.

The *Parshall flume* may be installed in canals and streams where the installation of a weir is unsuitable for lack of head. For various widths of the profile from 0.25 to 1.75 m, this flume measures discharges of up to 2.00 m^3/s, but larger flumes also exist. The flume may be built of cement or of wood, the latter being used for temporary measurements.

In all these measuring devices, a stage gauge must also be installed. Its distance from the weir must be equal to at least eight times the head.

The Parshall flume is used in places where the current is quiet and in places where the contraction of the stream is greatest.

On streams whose water surface is usually too irregular or for which the area of the cross-section cannot be satisfactorily measured — mountain brooks are good examples of this — or where the surface of the stream is completely frozen over, the dilution method may be used, as described later.

Measuring by a current meter

Very detailed instructions for such measurements exist in many countries. The following are based on the Czechoslovak instructions and the

Fig. 3.18. Stream-gauging station from Fig. 3.16

use of the METRA FB-1 current meter. Nevertheless, they do not differ from the internationally accepted standards and they permit the use of any other current meters, such as the OTT or Neyrpic instruments. These are all propeller-type current meters. The cup current meter (Price meter) is not used in Europe. For British practices, the reader is directed to British Standard 3680 and reports of the National Engineering Laboratory (Fluids) (Nos. 81, 72, 94, 65, 148, and others).

The principle of ascertaining the average cross-sectional velocity of flow by current meter measurement consists in dividing the cross-section into vertical bands for which an average vertical velocity is computed from point velocities measured by the current meters at characteristic depths on the vertical. This procedure involves an integration of velocities in space (over the area of the cross-section), since the velocity in each point of the cross-section is different. The velocity varies also in time; it pulsates at very short intervals. This pulsation is, however, integrated into an average by the very principle of the current meter.

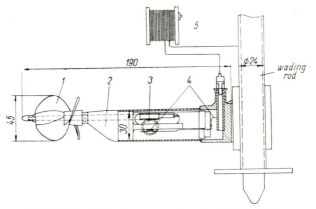

Fig. 3.19. Czechoslovak current meter FB1 — Metra

The instrument (*Fig. 3.19*) is composed of a propeller on an axle (1), the body (2), the revolution-counting device (3), and the electric contact (4). The whole current meter is suspended either on a rod or on a cable. A wire leads from the contact (4) to the buzzer (5), or any other signal- or revolution-counting device. The axle of the propeller of the METRA current meter is immersed in kerosene which fills the body of the meter. After fifty revolutions, a gear and eccentric cam make the contact and a signal is transmitted. The current goes through the rod or the suspension wire and a second insulated wire is needed to connect the buzzer with the meter. There are two buttons on the buzzer itself, one of which serves to connect the battery, the other to test the buzzer in a short-cut circuit.

The relation between the rotations n of the propeller and the velocity v

is indicated by the calibration graph of the propeller (*Fig. 3.19*) and its equation

$$v = \alpha + \beta n,$$

in which v = the velocity of current at a point,

α, β = constants indicated in a certificate provided by the current meter manufacturer.

Every current meter must be calibrated before being used in the field. This is done in the following way: the current meter is fixed on a rod or cable and towed through still water by an electrical carriage in a long rectangular or circular calibration tank at a fixed velocity. The number of revolutions is recorded and the velocity-revolution relation ascertained.

The Czechoslovak stream-gauging instructions recognize basic, simplified, and integration measurements. For practical needs on smaller rivers, only the simplified measurement may be used and only this method is described below.

The following material is needed for measuring shallow streams up to 1 m in depth and from 15 to 20 m wide (wading measurement):

(a) current meter with complete instrument kit including wading rod with base plate, wiring and signal or counter device (buzzer);

(b) stopwatch (unless included in the counter device of the meter);

(c) measuring survey tape (20 m), pegs and pins;

(d) levelling rod, or meter-and-sounding weight, portable staff gauge;

(e) spare battery, spare wire, electric plugs, pocket knife, flat pliers;

(f) blank forms and writing paper, pencils, etc.

The portable staff gauge is a staff about 1 m long, pointed at one end, and reinforced with metal edges, if possible.

For measuring deeper streams which are wider than 20 m, the following additional equipment will be needed (boat measurement):

A calibrated steel cable of 5 to 15 mm diameter, marked at every 1 m, and 100 to 150 m long, with a winch arrangement, 4 or 5 pieces of steel of 1 m, wooden or metal pegs and, in case of a swift stream, a tow cable of the same length as the calibrated cable.

If the measurement cannot be taken from a bridge, and if the stream is too deep for wading boots, a flat-bottomed boat or a special boat must be used. A platform mounted on two canoes (laminated, if possible) has proved very effective for measurements (*Fig. 3.20*) on wide but shallow streams. It is easy to handle and offers little resistance to the current, which makes it possible, in many cases, to use only the calibrated cable without a tow cable. Also, the canoes can be transported on top of a light pick-up truck or station-wagon car. This arrangement was designed and tested by the Civil Engineering Faculty, Prague Technical University.

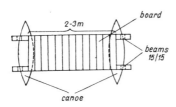

Fig. 3.20. Canoe gauging raft

Measuring procedure

The gauging site has been selected according to the principles given above. The cross-section is staked out by pegs or stakes perpendicular to the streamline with the aid of a pentaprism — a line may be staked out for this purpose along the bank in a straight reach of stream. A tape is stretched across the stream. In larger rivers, the cables are stretched across from a flat boat and the cable is anchored to the bank in such a way that the zero of the calibrated cable is on the left bank (looking downstream). If the boat is using the calibrated cable as a tow cable, care must be taken that it is quite taut. The width of the stream is measured in decimetres. After the tape or cables have been stretched, the current meter is assembled. It is screwed onto the rod and the signal device is tested for correct contacts; immediately before measuring, the METRA current meter is filled with kerosene.

In a narrow cross-section, the depth is measured simultaneously with the velocity. If there is a staff gauge at the site, the auxiliary staff gauge is situated between 100 and 200 m downstream, right next to the bank, so that the water surface slope may also be measured at the same time, as described below. If there is no staff gauge in the cross-section, the auxiliary one is situated directly at the site, to ascertain possible stage variation during measuring.

If the section is wide, it must first be sounded for depth of the stream by a wading rod, sounding weight, or a special fathometer.

The depth measurements should be located at regular distances and they should be closer together near the stream banks. Efforts must be made to register, as far as possible, all irregularities in the stream-bed cross-section. The number N_v of depth-measuring verticals is equal to or larger than the number N_z of velocity-measuring verticals.

For streams with a water surface up to 10 m wide, the depth verticals and the velocity verticals are the same and spaced at one metre intervals.

For wider streams, the numbers indicated in *Table 5* are used.

Table 5

Average depth of the stream H (m)	Number of velocity-measuring verticals								
	7	8	9	10	11	12	13	14	15
	to width of water surface B (m)								
0.5	7	10	20	30					
1.0		8	15	30	45				
1.5			12	25	40	70			
2.0				20	35	70	105		
3.0					30	60	100	140	
5.0						40	80	120	180

The number of velocity-measuring verticals is derived, in *Table 5*, on the basis of a preliminary estimate or measurement of the average depth and width of the stream surface. The distance calculated between the verticals $b_v = \dfrac{B}{N_v}$ is approximated to 1/5, 1/4, 1/2, 1, 3/2, 2, 5/2, 3, 4, 5 m.

In the simplified velocity measuring, the current meter is set in 3, 2, or 1 point of each vertical.

If 3 points are used, they are in 0.2, 0.4, and 0.8 of the depth.

If 2 points are used, they are in 0.2 and 0.8 of the depth.

If 1 point is used, it is in 0.4 of the depth.

(Measured from the bottom to the surface.)

If the ratio of the water-surface velocity to the average cross-sectional velocity is to be computed, the surface velocity is measured by the current meter directly under the surface, but the propeller must be entirely submerged.

Three or two points on each vertical are considered necessary for reliable discharge measurements. The points should not be more than 1 m vertically apart.

Only one point on the vertical is measured when pressed for time (during floods) or in a shallow bed and along the bank.

According to instructions, the FB1 Metra current meter should measure 400 revolutions (8 buzzing signals) at each point. But if the fourth signal (200 revolutions) sounds after only 45 s, the measuring at this point may be terminated. After the propeller is set at the point, it should be allowed to run through one signal (50 revolutions) before the stopwatch is started. The watch is usually started and stopped at the end of a signal which generally lasts several seconds. Intermediate signals are read with approximation to the second and the final signal is read with an accuracy of at least 0.2 s. Now, however, automatic revolution-counting devices, with incorporated clockwork, are used almost exclusively.

At the beginning and end of measuring, the water stage is measured by a staff gauge. If it fluctuates, it must be checked more often. If there has been a variation of 50 mm only, and in one direction, the average stage H_m is calculated as the arithmetic average of the initial and final stage.

If the fluctuation was more than 50 mm or if it was both ways (rise and fall), the average stage H_m is calculated by the equation

$$H_m = \frac{\Sigma v_{ms} H b}{\Sigma v_{ms} b},$$

where v_{ms} = the average velocity in a vertical,

H = water stage read when measuring this vertical,

b = half the distance (at the surface) between the measured and adjacent verticals.

The measurements are recorded into a printed form shown in *Table 6*.

The most common field troubles occur through malfunctioning of the signal system. It is therefore advisable to have a spare current meter or, at least, a spare signal system (buzzer and connecting wires). The malfunction results in irregular or fading signals. The buzzer, bell, or light should be checked to begin with. If they are in order, all contacts must be tested and the circuit checked.

Table 6

Station _____ Stream _____ Date _____

HYDROMETRIC RECORD

Measurement cross-section: good, fair, poor

Current meter No _____ on rod, on cable _____ Measured by: _____

Discharge _____ m_3/s _____ , cross-section area _____ m^2.

Average velocity _____ m/s, cross-section width _____ m.

Average depth _____ maximum depth _____

Number of measuring verticals: _____ of depth soundings: _____ of points

on vertical: _____ Water stage varied from _____ to _____ cm

(average _____ cm)

1	2	3	4	5	6	7	8	9
V	L	h	H	t = final time and inter-mediate times ;	V	V_{ms}	$V_{ms}h$	Notes
Number of vertical	Distance from bank	Total depth of the vertical	Depth of the point (from surface)		Velocity from the rating graph of the current meter	Average velocity in the vertical (computed)	Average velocity in the vertical multiplied by depth	Stages, time of beginning and end of measurement, etc.

Note: It is advisable to indicate also the weather, particularly the wind speed and its direction in relation to stream current, since the surface velocities are influenced by it.

Modern signal systems are based on an electronic counter of every revolution, with digital registration. A built-in stopwatch automatically registers the time. Thus, most of the difficulties described above may be avoided. Nevertheless, circuit trouble may occur in the most sophisticated signal system. While a simple signal system may be repaired on the spot, when using electronic counters, a spare is absolutely indispensable.

Two persons are the minimum needed for measuring by wading; one to hold the rod with the current meter and the other to measure time in the water and to take records. When measuring from a boat, an additional person is generally needed to navigate the boat.

Computation of discharge from current meter measurement

The record can be interpreted and computed by graphical and digital methods. Both are almost equally accurate, but the digital is faster.

Digital method used in Czechoslovakia

The measured times t from column 5 (*Table 6*) are checked and, in considering intermediate times, excessive deviations from the average and clear errors are eliminated. The final time measurement in each line is divided by the number of revolutions — use of a slide rule is recommended. From the rating graph or from the equation, or rating tables of the meter, the point velocity is read and entered in column 6 of the table. The average velocity in the verticals is computed by the following equations:

for 3-point measuring,

$$v_{ms} = \frac{1}{4}(v_{0.2} + 2v_{0.4} + v_{0.8})$$

for 2-point measuring,

$$v_{ms} = \frac{1}{2}(v_{0.2} + v_{0.8})$$

for 1-point measuring,

$$v_{ms} = v_{0.4}.$$

The v_{ms} thus computed is entered into column 7 of the table and multiplied by the depth of the vertical. The result $v_{ms}h$ is entered in column 8. This computation must be revised in the case of the first and last verticals, if their distance from the bank is different from the half distance of verticals in the stream $\frac{1}{2}b_v$. If this distance is equal to $\frac{1}{2}b_v$, no special revision is necessary. *Fig. 3.22* represents a case in which the distance of vertical I from the bank is less than $\frac{1}{2}b_v$. In this case, instead of h

in column 3, the value of $h_0 = \dfrac{F_0}{b_0}$ is entered (in which F_0 is the area of the cross-section at a distance b_0 from the bank — it is always advisable to sketch the area F_0). Into column 7, instead of v_{ms}, enter $\dfrac{2}{3} v_0$, where v_0 is the maximum velocity in a vertical in the band of width b_0. This velocity is computed from a graph of the average vertical velocities, which is constructed on graph paper, as indicated in *Fig. 3.22*. Columns 2 and 7 of *Table 6* are used for the plotting of the graph. This graph is always annexed to the record. The bank correction of average vertical velocity is indicated in the graph by cross-hatching the corresponding bank area $\left(\text{from bank until } \dfrac{1}{2} b_v \text{ distance from the first — or last — vertical}\right)$.

All entries in Table 6 are in m and m/s. The total of column 3 (depths), multiplied by b_v, is the area of the cross-section A. The total of column 8 ($\Sigma v_{ms} h_i$), multiplied by b_v, is the discharge Q. Thus, average velocity in the cross-section (m/s) is the ratio of totals from columns 3 and 8. The following equations are the mathematical expression of the above operations.

$$A = b_v \, \Sigma h_i,$$

$$Q = b_v \sum_0^B v_{ms} h_i,$$

where $B =$ the width of the stream,

$$v_{\text{average}} = \frac{Q}{A}.$$

The above computation is based on the assumption that all the distances between the measuring verticals (with the exception of possible bank corrections) are equal.

Problem 2

Computation of discharge from field records
(in accordance with the instructions of the Czechoslovak Hydrological Service)

The water stage H of a river $= 1040$ mm corresponds to an average depth $h = 1$ m (estimated on the basis of previous measurements). The width of the cross-section at surface is 51 m. The hydrometric site is good for discharge measuring. According to *Table 5*, 18 depth-sounding and 9 measuring verticals were selected. The distance between the depth-sounding verticals is adjusted to 3 m; the distance between measuring verticals b_v is therefore 6 m, except for vertical IX, which is only 4 m from vertical VIII. A current meter METRA FB 1 No. 5480049 with a constant $\alpha = 0.014\ 38$ and $b = 0.117\ 21$ was used. At each measuring vertical, two points were measured at 0.2 and 0.8 of depth. The results were recorded in the record book (*see Fig. 3.22*). It may be pointed out that, at vertical I, measuring at a depth of 140 mm, was terminated after four signals because the total time was more than 45 s. A similar remark applies to vertical IX.

Since the distance between the stream banks and the first and last measuring verticals did not equal $\frac{1}{2} b_v$, it was necessary to apply bank correction. In place of the measured depth $h = \mathbf{36\ cm}$ at vertical I, $h_0 = \dfrac{F_0}{3} = \dfrac{33}{3} = \mathbf{11\ cm}$ was entered in column 3, and for vertical IX, instead of $h = \mathbf{310\ mm}$, $h_0 = \dfrac{24}{3} = \mathbf{8\ cm}$ was entered.

The velocity graph (*see Fig. 3.21*) was constructed from average vertical velocities. The velocities at sounding verticals were read from this graph and entered in column 7 of *Fig. 3.22*. Thus, not only velocities at measuring verticals were used, but also all sounding verticals for computation of discharge.

A bank correction was made for the measuring verticals I and IX. $\frac{2}{3} v_0$ at vertical I $= \frac{2}{3}\mathbf{0.420} = \mathbf{0.280\ m/s};$ at vertical IX, $\frac{2}{3} v_0 =$ $= \frac{2}{3}\mathbf{0.139} = \mathbf{0.093\ m/s};$ v_0 was read from the graph (*Fig. 3.21*) plotted from column 7 of *Fig. 3.22*.

The total of column 3 $= \mathbf{1881\ cm} = \mathbf{18.81\ m}$. Since $b_v = 3$ m, the area of the cross-section $\mathbf{A}$ will be:

$$A = \Sigma h b_v = 18.81\ \text{m} \times 3\ \text{m} = 56.43\ \text{m}^2.$$

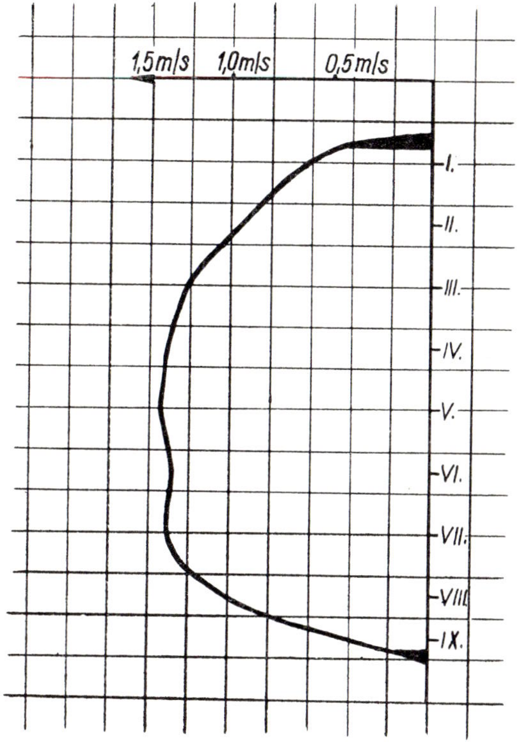

1	2	3	4	5				6	7	8	9
Z	L	h	H	t				v	v_{mz}	$v_{mz} \cdot h$	
4,20	θ		–						0,280		14ʰ35 / 104 cm
5	11,36 / 70		14 / 56	11 24 36 48 / 7 15 23 90					0,421 0,625	310 / 4370	
I. 7	84		–	38 49 55				0,503 0,638 0,747		6700	
10	98		20 / 78	44 53 62 / 30 25 28				0,686 0,896 0,941	0,797 0,941	9220	
II. 13	101		–	5 11 18 23 / 34 42 47				0,932 1,517	1,002 1,884	11020 / 12600	
16	103		21 / 82	4 8 12 16 / 20 13 27				1,038 1,582	1,276 1,310	13120 / 14280	
III. 19	109		22 / 87	6 12 18 25 / 30 35 41				1,152 1,547	1,332 1,349	14800 / 15910	104 cm
22	111		–	7 11 15 / 19 22 26				1,000 1,498	1,322 1,294	15720 / 15400	
IV. 25	118		24 / 94	4 10 15 21 / 25 31 36				1,115 1,527	1,309 1,321	15700 / 15580	
28	119		–	8 12 16 / 20 23 27				1,090	1,288 1,033	15700 / 12820	
V. 31	119		24 / 95	5 11 16 21 / 26 32 38				0,801 1,266	0,780 0,406	9820 / 4510	
34	120		–	4 8 18 16 / 20 34 28				0,360 0,452	0,490	70	104 cm
VI. 37	118		24 / 94	5 10 15 24 / 32 37					0,093		16ʰ02
40	128		–	4 7 11 15 / 19 23 27							
VII. 43	124		25 / 99	7 15 24 32 39 / 45 53 59							
46	126		–	5 10 14 18 / 23 28 32							
VIII. 49	111		22 / 89	22 77 33 52 68							
51	8,91		13 27 40 57 58								
IX. 53	θ		–							207653 cm²/s	
55	1881,3									62,296 m³/s	
55,20	5643 m²										

Fig. 3.22. Table for discharge computation (see also Fig. 3.21)

Fig. 3.21. Discharge computation (see also Fig. 3.22)

The total of column 8 is **20.765 m** (the notation in *Fig. 3.22* is in whole numbers, e.g. in cm²/s). Thus

$$\Sigma v_{ms} h = 20.765$$

and

$$Q = b_v \Sigma v_{ms} \cdot h = 20.765 \times 3 = 62.295 \text{ m}^3/\text{s} = 62.3 \text{ m}^3/\text{s}.$$

Average cross-section velocity

$$v_{av} = \frac{Q}{A} = \frac{62.295}{56.43} = 1.1 \text{ m/s}.$$

Results

At a water stage $h = $ **104 cm** and a cross-section area $A = $ **56.43 m²**, the discharge was $Q = $ **62.3 m³/s**. The average velocity is **1.1 m/s**.

Many accidental factors, interrelated and independent, affect the accuracy of measuring. These include, for instance, the error in the meter rating, and thus of the meter equation, the pulsation of the point velocities of the water, the number and location of the measuring verticals, number and location of points on the verticals, inaccuracies in sounding (error in the area of the cross-section), weather influence, subjective personal factors caused by mistakes of persons measuring, etc.

The accuracy of the results may be ascertained by the relative accuracy

$$p = \pm \frac{S}{Q_{com}} 100 \text{ per cent}$$ where $S = $ the standard deviation computed from all discharge gaugings with the same instrument in a long-term systematic measuring at the cross-section; $Q_{com} = $ the discharge computed from the measurement under consideration (for standard deviation computation *see page 132*). The relative accuracy p per cent may also be computed as a function of the factors influencing the stream gauging given below and valid for a specific current meter. The accuracy standards for systematic long-term measuring have been set at ± 4 per cent with a reliability of 90 per cent. This means that with a long series of discharge measurements, 90 per cent of them should have an accuracy of ± 4 per cent or better. The remaining 10 per cent have an accuracy of less than 4 per cent but this should never be below 8 per cent.

In order to attain such accuracy, the following rules must be observed:

(a) the points on the verticals should always be at the same relative depth;

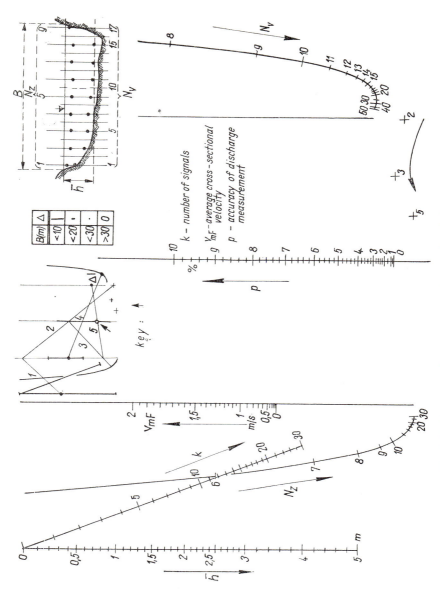

Fig. 3.23. Nomogram for accuracy evaluation for discharge measurements

(b) the sounding verticals should be uniformly distributed in the cross-section—there should be more verticals near the bank in order to reflect the shape of the cross-section properly;

(c) the velocity in the individual points should be measured with the same number of meter revolutions (with the exception of low velocities).

The ratio between the main factors influencing the accuracy of measurements with a 90 per cent reliability is given by the following equation:

$$\frac{p^2}{1.645^2} = \frac{0.36 + 186R^{0.8}\frac{1}{k}}{N_z f_1(a)} + \frac{f_2(a)}{N_z} + \frac{14.4 \times 10^4}{N_z^4} + \frac{4.84 v_{av}^2}{N_v} + \\ + \frac{6.25 \times 10^4}{N_v^4},$$

where p = the accuracy ($\pm$) of the computed discharge data [per cent],

R = average depth of the measured cross-section [m],

v_{av} = average velocity [m/s],

k = number of signals from Metra FB 1 meter (every 50 revolutions),

N_z = number of measuring verticals,

N_v = number of sounding verticals,

$f_1(a)$, $f_2(a)$ = function of number of measuring points on vertical.

For two-point measurement $f_1(a) = 2$, $f_2(a) = 1$.

For one-point measurement $f_1(a) = 1$, $f_2(a) = 6.25$.

The solving of this equation is facilitated by the nomogram (*Fig. 3.23*) which also includes the influence of the width of the stream B. The above equation was derived by a multiple correlation of standard deviation to the factors influencing the gauging.

It should be pointed out that both the nomogram and the equation numerical parameters refer only to current meter Metra FB 1. They can, however, be ascertained for any current meter.

Measuring velocity by a float

Basically different, but similar in practice, is the method for measuring average velocity by a float. This method is used if there is no current meter available and also during floods, when it is difficult or even impossible to measure with a meter. It is also used where it is not necessary to determine the exact discharge or velocity. However, if float measuring is carefully executed, it differs no more than 15 per cent from measurements made with a current meter.

Any sort of floating object may be used as a float, but it should not protrude out of the water too high and it should not be too light. If measurements are being taken in a straight, deep and regular channel, a rod travelling vertically may be used to measure an average velocity in the vertical.

The float may be a wooden cross, for instance, to which a wire weight has been attached, or else a bottle with a rod inserted.

The following material is needed for measuring discharge with a float:

(a) the necessary number of floats (a large number is recommended);
(b) a stopwatch or a watch with a second hand;
(c) measuring tape and marking pegs (6);
(d) levelling or sounding rod, auxiliary staff gauge.

If the stream is deep, a flat-bottomed boat or any other type of boat is necessary. If velocity only is to be measured and if the area of the cross-section is already known from previous measurements (during floods), depth-sounding is not necessary.

A straight, uniform reach of stream is chosen, three to five times longer than the width of the stream.

A straight line is staked out on the bank, parallel to the stream, and on it three points are marked, the middle being the measuring cross-section. The beginning of the measuring reach is upstream (at a distance of two or three times the width of the stream) and the end is downstream (at a distance of one to two widths of the stream) (*see Fig. 3.24a*). The cross-sections are marked perpendicularly to the stream. For rivers less than 20 m wide, the depth is ascertained every metre; for wider streams, it is ascertained in accordance with *Table 5*. An auxiliary staff gauge is set, if there is not already a permanent one, and observed while sounding. After the area of the cross-section is measured, the floats are thrown one after another, about 10 to 15 m upstream from the beginning of the measuring reach. Three floats are used for streams 20 m wide and, for wider streams, they are placed every 10 m across the width of the stream. One float is sufficient for streams narrower than 5 m.

An observer stands at the beginning of the measuring reach. The moment when the float reaches the beginning of the reach is signalled to the observer at the end of the reach (by waving of arms, blowing a whistle, etc.). When the float reaches the end of the measuring reach, the end observer registers the time and signals for another float to be launched. Several floats may be thrown in at once if experienced observers are at hand. In small streams, when measuring with one float, a single person may perform the entire measurement (*Fig. 3.24b*).

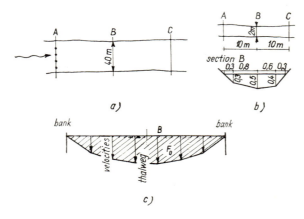

a)   b)

c)

Fig. 3.24a, b, c. Stream gauging by float (area-velocity method)

The average surface velocity v_s is then computed graphically from the velocity of the individual floats by plotting a velocity graph (*Fig. 3.24c*) and by measuring its area A_0. Then $v_s = \dfrac{A_0}{B}$, where $B =$ width of stream. If only one float is used, the measured velocity is the average surface velocity. From this velocity, the average cross-sectional velocity is computed with the greatest accuracy if, from previous current meter measurements, the ratio of both velocities was computed. This ratio d varies between 0.85 and 0.96 and is usually quite steady at different water stages. It is also possible to establish d as the function of H (stage) and plot a curve $d = f(H)$.

For small streams, with measurements by only one float, the following empirical equation may be used:

$$d = \frac{v_s + 2.354}{v_s + 3.129} \cdot$$

The discharge is then computed by area-velocity method

$$Q = Av_{av}.$$

Measuring of the slope of the water surface is also part of basic hydrometric work. This is often overlooked, although it is the most important element of stage-discharge relations and is often absolutely necessary for a proper plotting of the discharge rating curve. The tendency and slope of the surface may differ with the same water stage, particularly if the surface is rapidly moving — rising or falling — as is usually the case during floods. The water surface of a natural stream is never a perfectly straight, inclined plane but, in view of the various degrees of roughness of the channel, it is composed of a number of distorted planes. In addition, the surface is not horizontal in the cross-section, especially in meandering reaches. It is lower along convex banks and higher along concave banks. Even though the differences are generally small, they can be significant in computing the slope of the water surface on shorter reaches of the stream. For more accurate measuring, measurements along both banks are thus necessary.

The measuring is done by staff gauges situated at a distance approximately equal to five widths of the stream. If only the water surface slope at the hydrometric station is to be measured, the staff gauges are situated upstream and downstream at distances indicated above. The readings of all the staff gauges must be simultaneous.

The zero of all the auxiliary staff gauges is the same. If staff gauges are not available, stakes can be used, hammered into the stream next to the bank and levelled. The water stage is then measured with an ordinary pocket ruler from the top of the stakes. Again, measurements must be taken simultaneously.

Very often it is necessary to determine the slope of the surface during the crest of a flood. For this, the straightest possible stretch of stream should be chosen with the most regular channel and flood plain and with the most homogeneous vegetation in the flooded sections. High-water marks are ascertained on trees and buildings, at a distance equal to at least five times the width of the water surface at the time of flood. All such marks are levelled.

In levelling, the greatest accuracy is recommended, because the slope of water surfaces is very small.

Often, special cableways are constructed across the stream for discharge measurements. Measuring from a cableway is no different from a bridge with a cable (wire) suspension of the current meter. A vertical and horizontal angular correction must be applied to the measured velocities. The depth measurement is made mostly with sounding weights.

The operation of the cableway and measurement from it require special training of the personnel. In most cases, such cableways are installed by specialized companies (one world-wide specialist is the Swiss company Von Roll). *Figure 3.25* shows a photograph of a cableway in operation.

Measuring flow under ice is a very particular type of discharge measurement. The ice thickness must be measured in addition to all the measurement procedures described above. Safety of personnel must be particularly heeded during such measurements.

The flow under ice may be measured by the dilution method, instead of with a current meter. This method has lately been perfected in France (Grenoble).

Measuring discharge by dilution method

The ISO Recommendation R. 555 includes a detailed description of this method. As also described in the WMO Guide to Hydrometeorological Practices, it is recommended only for sites where conventional measuring cannot be employed: shallow depths, extremely high velocities, excessive turbulence. The injection of a tracer in the stream — the dilution method is based on this principle — may either be at constant rate (plateau gauging) or instantaneous (gulp method).

A solution of a stable or radioactive chemical is injected into the stream at either a constant rate or all at once. The solution will be diluted by the

Fig. 3.25. Stream-gauging cableway

discharge of the stream. Measurement of the rate of injection, the concentration of the tracer in the injected solution, and the concentration of the tracer at a cross-section downstream from the injection point (sampling section) permits the computation of the stream discharge. The accuracy of the method critically depends upon two factors.

(a) Complete mixing of the injected solution throughout the stream cross-section before the sampling section is reached. If the tracer solution is continuously injected, the concentration of the tracer should be constant throughout the measuring section. If the tracer is all injected at once, $\int_0^T c\, dt$ should be the same at all points in the section, where c is the concentration and T is the time in which all of the tracer passes a particular point in the section. These criteria should be verified before the location of a measuring section is finally chosen. As a general guide, only the required length L between the injection and sampling section will be:

$$L = 0.13C\,\frac{(0.7C + 6)}{g}\,\frac{b^2}{h} \qquad \text{(metric units)},$$

where b = the average width of the wetted cross-section,
       h = the average depth of flow,
      C = the Chézy coefficient for the reach ($15 < C < 50$),
      g = gravitational constant.

(b) No absorption of the added tracer by stream bottom materials, sediments, plants, or organisms, and no decomposition of the added tracer in water of the stream. The concentration should be determined at the sampling section and at least one other cross-section downstream to verify that there is no systematic difference in the mean concentration from one sampling section to another.

Any substance may be used as a tracer if it:

dissolves readily in the stream water at ordinary temperature;
is absent in the water of the stream or present only in very small quantities;
is not decomposed in the stream's water and is not retained or absorbed by sediments, plants, or organisms;
can be detected (concentration) by simple methods;
is harmless to man and animals in the concentration it assumes in the stream;
is not expensive.

The cheapest tracer is common salt. Where the tracer is instantaneously injected into the stream, the required quantity is not particularly large and detection by conductivity methods is relatively simple.

Sodium dichromate is used extensively in the dilution method. Its solubility in water is relatively high (1.85 kg per kg of water at 0 °C and 2.06 kg per kg of water at 20 °C), and the salt satisfies most of the requirements stated above. Colorimetric analysis permits the measurement of very low concentrations of sodium dichromate.

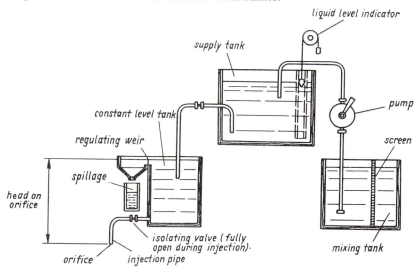

Fig. 3.26. Stream-gauging by dilution method (instrumentation)

Radioactive elements such as gold 198 and sodium 24 have been used in the dilution method. Concentrations of these elements as low as 10^{-9} may be accurately determined with a counter or count-rate meter with the sensing probe suspended in the stream or in a standard counting tank. Although radioactive elements are ideal tracers for the dilution method, the health hazards may limit their use in the measurement of stream discharge in some localities. For the use of radioactive tracers, see details in the Guide published by the IAEA (see references). *Fig. 3.26* shows a schematic diagram of the instrumentation for injection of the tracer, as illustrated in ISO Recommendation 555.

The equations used to compute the stream discharge Q are based on the principle of continuity of the tracer.

$$Q = q \frac{C}{c} \qquad \text{(continuous injection)},$$

$$Q = \int_0^\infty \frac{CV}{c} \, dt \qquad \text{(sudden injection)},$$

where q = rate of injection,
$\quad C$ = concentration of injection solution,
$\quad c$ = concentration in the stream at the sampling section,
$\quad V$ = volume of injected solution,
$\quad t$ = time.

Fig. 3.27 gives the diagrams of the tracer's dilution (wave) for the *'plateau'* and the *'gulp'* methods respectively.

If the general requirements of the method are met, the accuracy of the discharge measurement depends on the measurement of C relative to c. This measurement can normally be made to an accuracy of one per cent, if a sample of the injected solution is diluted with the river water by a known amount so that $C = c$ and both concentrations are then measured in exactly the same manner.

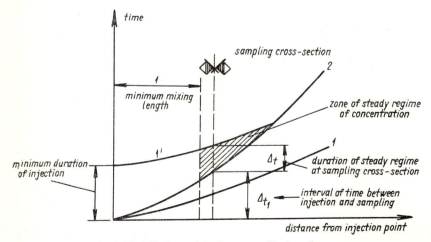

Fig. 3.27. Dilution method—tracer dilution diagram

Fig. 3.28. Bed-load sampler of the Hydraulic Research Institute in Prague

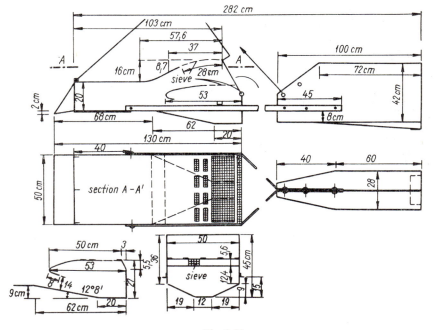

Fig. 3.29

Measurement of sediment transport

Instruments for measuring bedload and suspended sediment have not yet been unified or normalized, either in Czechoslovakia or elsewhere. A wide-necked bottle (such as a milk bottle) with a cork with two holes is used for suspended sediment sampling in Czechoslovakia. Water and sediment enter the bottle through the lower aperture and air comes out through the top. Samples are taken from various places and at various depths.

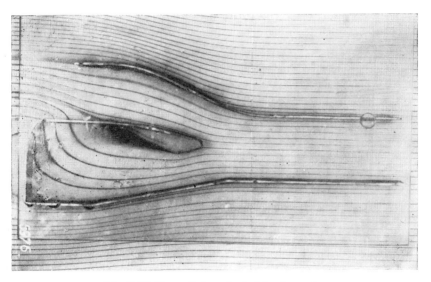

Fig. 3.30. Streamlining of a bed-load sampler

Bedload is collected in samplers consisting of a wire basket or box which is dragged at the bottom of the stream. *Fig. 3.28* shows the sampler of the Hydraulic Research Institute in Prague; *Fig. 3.29* is the diagram of this sampler and *Fig. 3.30* shows the streamlining of a sampler.

The methods used for sampling sediment and bedload are extremely divergent and still in the research stage. Also, as such measurements are mostly within the responsibility of highly specialized personnel, the reader is directed to the Guide to Hydrometeorological Practices of WMO, as well as to national publications (Ven Te Chow, 1965). A British Standard for sediment samplers and bedload measurement is in preparation.

(d) Field hydrological survey

Before hydrological data for projects are computed, a hydrological field survey must be made. For irrigation and drainage projects, it may be part of the geodetic survey. This survey is usually divided in two parts. The first is preliminary and serves to ascertain the hydrologic factors of the catchment and of the stream itself. It does not result in detailed, quantitative data. A notebook and pencil are sufficient for this part of the survey.

The start is generally at the lowest cross-section downstream (at the mouth of the stream) and the survey develops upstream.

Before starting the survey, all available material about the basin and the stream should be studied. This includes topographic maps (1 : 25 000, 1 : 50 000, or, for more detailed studies, 1 : 1000 and 1 : 5000), soil and geological information (soil and geological maps), climatic data including locations of meteorological stations, economic data including plans for watershed management, and stream-water exploitation (for farming, industry, water power). It is very important to locate and examine all hydraulic structures on the stream — weirs, mills, pumping stations and canal intakes, as well as inflows of waste water from sewers and industrial plants.

During the preliminary survey, the stream is followed, the character of the different cross-sections being noted (their selection for gauging was described above), checking the facts contained in previous documents. In representative sites selected previously on the map, the character of forests and vegetation, and the soil composition, is ascertained. At the same time, flood marks are looked for, usually on buildings and bridges, or on trees if necessary. The location of such marks is noted in the map and included in the log book so they can be found by others. Accessible and possibly little-used wells are also located and noted, so that fluctuations of groundwater level may be measured.

The preliminary survey is intended to ascertain the general character of the stream and basin. If there have been no direct measurements of flow on the stream, on the basis of preliminary survey, an analogous basin may be found, for instance, on the same stream, upstream or downstream.

A detailed survey has a more precisely defined purpose: it is undertaken with surveying or even hydrometric equipment. This includes a surveying outfit (1 theodolite, 1 levelling instrument, levelling staffs, measuring tapes, prism, etc.). A pocket levelling instrument with rod and tape is often satisfactory. The hydrometric instruments were described above. In addition, it is good to take along a sounding weight for measuring the depth of the ground water in wells in the area under study.

A detailed survey is usually already directed toward a certain purpose: it has to ascertain the peak discharges of past floods from high-water marks, to survey the area and volumes of the flood plain for flood routing computation or flooded area and volumes of a proposed reservoir.

If a very accurate topographic map on a 1 : 1000 scale, for instance, is needed, the necessary survey is executed by geodetic surveyors who prepare a full report for the hydrologists.

In a detailed survey, the peak stages of past floods and the corresponding cross-sections of the stream and the valley are surveyed. Since all the marks were located and evaluated during the preliminary survey, only those which are most reliable and clear are levelled.

Often the velocity of the flowing water is measured, to give some indications of time elements for concentration or lag time evaluation in hydrograph analysis and synthesis. Often only levelling work is done, in which the elevations of weir crests, of outlets, canals, intakes, are levelled. It is evident that an engineer-hydrologist must usually be well acquainted with at least basic geodetic surveying methods. A good field survey may bring considerably more precision to further hydrologic computations and analysis, and be of great economic benefit to projects. One rough but direct measurement is better than dozens of indirect computations, accurate as they may be. This saying of the phycisists of the last century is still more than valid for today's hydrologists.

4 Hydrological Analysis and Design

4.1 Accuracy of computations in hydrology

The accuracy of computations must correspond to the accuracy of observations and measurements. There would be no purpose in selecting sophisticated and precise methods of analysis if the observed data would not approximate the accuracy of computations. An example is calculating the average precipitation over a basin. If there are no major differences between observations of individual stations, it is unnecessary to use the precise, but time-consuming, isohyetal method which yields results which are accurate to a tenth of a millimetre, if the precipitations are measured in whole millimetres. *Table 7* indicates the accuracy of computations as recommended in USSR by A. A. Lutcheva (1950).

In rounding off the size of the area of a basin, discharge and amount of flow, the following principles are recommended:

$$
\begin{aligned}
&< \quad\;\; 1 \text{ round off to } 0.001, \\
&1 - \quad 10 \text{ round off to } 0.01, \\
&10 - \;\; 100 \text{ round off to } 0.1, \\
&100 - 1000 \text{ round off to } 1, \\
&> 1000 \text{ round off to } 10.
\end{aligned}
$$

The following rule should be used in rounding off figures: if the deleted digit is less than 5, the last digit should remain unchanged. If it is 5, the last digit should be rounded up if it is an odd number; if it is even, it is left unchanged. If the deleted digit is more than 6, it should be rounded upwards. For example: 0.334 becomes 0.33; 0.335 and 0.345 both become 0.34; 0.346 becomes 0.35.

Table 7

Element		Accuracy	Element	Accuracy
Time		1 s	Long-term average runoff	Two- or three-figure numerals
Depending on the stream size	Water depth	1, 10, 20 cm	Water and air temperature	0.1 °C
	Water stage	1 cm	Runoff depth	1 mm
	River width	0.1, 0.5, 1 m	Humidity; saturation deficit	1 mm Hg
	Stream length	0.1, 1 km	Precipitation depth	1 mm
Snow cover depth		1 cm	Runoff coefficient	Two-figure numeral
Slope (head)		0.1 per cent	Evaporation	1 mm
Velocity of flow		0.01 m/s	Wind speed	0.1 m/s

The choice of scale for graphs depends on the accuracy of the maps used or the purpose.

The following scales are usually used for transverse and longitudinal cross-sections:

(a) *horizontal scale*

1 : 50	500	5 000
1 : 100	1 000	10 000
1 : 200	20 000	20 000

(b) *vertical scale* is usually 10 to 20 times more compressed than the horizontal scale.

The following scales are used for the hydrograph and water stage graphs (annual):

(a) *horizontal scale* (time)

$$1 \text{ cm} = 10 - 20 \text{ days}$$

(b) *vertical scale*

water stage 1 cm = 10, 20, 50, 100, 200 cm,
discharge   1 cm = 10, 20, 50, 100, 200 m³/s.

4.2 Methods of hydrological calculation

Data and observations from a network, supplied to the hydrologist and supplemented by his own field surveys, cannot usually be used directly as design data for projects. They have to be processed by engineering hydrology methods and expressed numerically or graphically.

There are two extreme alternatives of availability of raw data.

Either sufficiently detailed and long-term series of directly measured data, yielded by measurement and observation, exist for the basin and on the site of the stream concerned or in its immediate vicinity. Alternatively, no direct observations on the stream exist at the concerned site or in its immediate vicinity.

There are many variations of these two basic alternatives, the three most common being as follows.

(a) There has been direct and sufficiently long stream-gauging on the stream, but at some distance from the site in question; this is the case in upper reaches of streams with a number of hydrometric stations on the lower reaches.

(b) No stream-gaugings have been made on the stream in question, but on another stream, of which the first is the main tributary.

(c) There are no stream-gaugings on the stream in question, but there have been sufficient stream-gaugings in its vicinity on another stream of the same character with similar basin characteristics (e.g. two tributaries of the same main stream, or representative basins).

It is primarily larger streams which had long-term stream-gauging (in Czechoslovakia these are basins larger than $100-150 \text{ km}^2$). Long-term gaugings on smaller streams are very scarce, even in countries with long hydrometric traditions.

In *chapter 1 on pages 9 – 10*, hydrological characteristics which are needed in the design of water resources and other projects are indicated. In economically evaluating these projects, the designer has to select such characteristics, which, after a cost-benefit study, will produce the most economical solution of the project. Such evaluation necessarily involves the consideration of risk, which must be qualified by its probability.

Consequently, the hydrological characteristics will have to be related to their probability of occurrence. Hence, the methods based on mathematical statistics and probability calculus will play an important role in hydrological analysis and processing of data.

These methods are used on a routine basis in Czechoslovakia and in many countries of central and eastern Europe, particularly in the USSR, and corresponding national standards, for design floods in particular, have been issued by government authorities. They will be referred to further on.

Nevertheless, probability evaluation and statistical approaches often present practical difficulties in their application, particularly in the absence of long-term series of data. In addition, they must be used with caution, since several theoretical reservations may be formulated with respect to their indiscriminate application.

Although modern developments of the so-called 'stochastic' processes, analysis and synthesis, resulting in simulation of time-space series of data, have lately partially alleviated the severe reservations of many hydrologists towards the probability approach, the difficulties of their use in the absence of reliable time series of data remain. Thus, an approach of physical analysis of the hydrological process and its factors, represented in part by empirical formulae and resulting in the so-called 'parametric hydrology', or 'deterministic' models, including conceptual computer simulation models, must be used in conjunction and must blend with the probability approach. A combination of the two approaches, whenever feasible, guarantees the best result in the determination of economically qualified and scientifically founded hydrological characteristics. Such a combination, when appropriate, will be used in this work.

4.3   Basic processing of time series, correlation, frequency distribution

Most of the raw data of hydrometeorological and hydrometric statistics supplied by networks is in the form of *time series*. Since the data necessarily cover only a certain period of observation, extended as it may be, such data represent only an observed sample of the entire *population* of the continuous process. While the time or space element of hydrological phenomena may be the same (one catchment observed for a certain period of time), each phenomenon is characterized by a certain value which varies in time and space. This characterization is called *variable* and its particular value is a *variate*. Thus, discharge or rainfall depth are variables. The variable may be continuous or discrete. The above variables are continuous. Different sizes of rocks contained in a bedload sample are a discrete variable.

If the value of one variate is independent of any other, the variable in question is a *random* variable. In hydrology, practically speaking, most processes are random processes and the respective variables are equally random. The random character of hydrological variables is most important for justifying the use of statistical methods for their processing.

All the time series (as well as other series) may be characterized by *statistical parameters.*

The simplest parameters characterizing a series are *means*. The arithmetic mean is indicated in hydrological characteristics in *chapter 1* of this book by the symbol *Mx*. Nevertheless, to comply with accepted statistical symbols, in general considerations of a statistical nature, it will be indicated by $\bar{x}$.

The arithmetic mean is computed as

$$\bar{x} = \frac{1}{n} \sum_{i=1}^{i=n} x_i,$$ (4.1)

where x_i = variate,

n = total number of variates in the series.

The mean does not indicate the dispersion of variates in the series. The simplest parameter of dispersion of a series is the *mean deviation d*

$$d = \frac{\Sigma(x_i - \bar{x})}{n}.$$

The most used parameter of dispersion is the standard deviation

$$d = \frac{\sum_{i=1}^{i=n}(x_i - \bar{x})^2}{n}, \tag{4.2}$$

for the entire population (very long series) or

$$d = \frac{\sum_{i=1}^{i=r}(x_1 - \bar{x})^2}{n - 1}, \tag{4.3}$$

for a sample (shorter series).

The ratio of a variate to the arithmetic mean is the reduced variate k_i known in eastern Europe, as the *module coefficient*

$$k_i = \frac{x_i}{\bar{x}}.$$

The difference between the maximum and the minimum values of the variable in a series is the *amplitude a*

$$a = |x_{max} - x_{min}|.$$

A further characteristic of a series is the frequency distribution of the values of the variable and is obtained by frequency analysis of the series, which is described later on. As has already been mentioned above, a frequent method in the computation of hydrological characteristics of a catchment is to adapt series of observations of one phenomenon to another phenomenon of the same catchment, or from another similar catchment which has more complete or more extensive observation data. An objective mathematical method of relating the series of observations in one catchment or between catchments is therefore needed. Such methods are either graphical or numerical, based on a statistical approach.

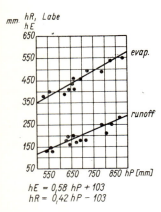

$$hE = 0,58 \, hP + 103$$
$$hR = 0,42 \, hP - 103$$

Fig. 4.1. Runoff-rainfall relation

(a) Graphical representations of the relation of series

This graphical relation is usually plotted in an orthogonal co-ordinate system with either arithmetic scales on both axes of co-ordinates or with special scales, usually logarithmic or probability scales, on one or both axes.

For instance, in the orthogonal co-ordinate system, with an arithmetic scale on the abscissa, it is possible to plot the annual precipitation hP and, on the ordinate, the average annual runoff depth hR and average total evaporation depth hE. For the same year (*Fig. 4.1*), these points concentrate around a straight line which expresses the ratio between hP and hR (the Czechoslovak Hydrological Service has termed this ratio the runoff law). The equation of this line will be

$$y = ax + b,$$

where $y = hR$,
$\qquad x = hP$.

If we determine a and b of this equation, we can calculate hR for an arbitrary hP, within the amplitude of the observed values. Beyond this amplitude it may be necessary to extrapolate a series of observations by another method of mathematical statistics. If the annual precipitation has been observed for a longer period than stream flow, the annual runoff yield for unobserved years may be roughly estimated from the graph.

Naturally, such estimates are very approximate, since the runoff is not an unequivocal function of precipitation, but is the result of the influence of many other factors, as indicated in *chapter 1*. The plotted points are therefore more or less dispersed around the straight line, called the line of best fit. The relation between runoff and precipitation is statistical, or correlational, and it may also be ascertained by means of correlational analysis.

A functional relation between two phenomena can be accurately expressed by an analytical function, that is, a certain value of one phenomenon inevitably corresponds to a single value of the second phenomenon.

Such relations are quite exceptional in hydrology: they are encountered rather in hydraulics. Such, for instance, is the relation between the head on a weir and the discharge Q, given on *page 99* and called the rating curve of the weir.

In measuring discharge in an open channel cross-section, a similar rating curve (relation between discharge Q and water stage H) is derived. This curve is, however, only fitted to a distribution of plotted points. The fitting of such a curve is illustrated in the following example.

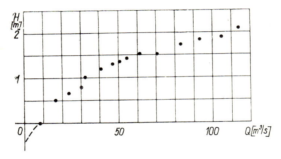

Fig. 4.2a. Stage-discharge curve in arithmetic scale

Problem 3

Values H and Q [m and m³/s], obtained from direct stream-gauging by area-velocity method with velocities measured by a current meter are in *Table 8*.

Table 8

H	0.00	0.49	0.78	0.88	1.03	1.20	1.28	1.34
Q	9.8	15.9	23.1	29.8	31.8	40.2	46.2	49.8

H	1.36	1.51	1.55	1.76	1.86	1.92	2.10
Q	54.5	61.0	71.0	82.5	93.2	105.1	113.4

From the table and from the graphical plotting (*Fig. 4.2a*), it is evident that the increase in discharge with rising water stage is not linear. The points concentrate around a curve which is exponential, and may be a parabola, usually of second degree.

The general analytical expression of an exponential curve is

$$y = ax^n \tag{4.4}$$

or

$$y = ax^n + b. \tag{4.5}$$

An exponential curve may be represented as a straight line in logarithmic co-ordinates. Numerically it can be expressed by taking logarithms of equation (4.4)

$$\log y = \log a + n \log x \tag{4.6}$$

and if

$$\log y = Y,$$
$$\log x = X,$$
$$\log a = A,$$

eq. (4.5) is linearized as

$$Y = nX + A.$$

This is often the point of departure in seeking relations in hydrological practice. If the relation between two phenomena is plotted in logarithmic co-ordinates and if the plotted points concentrate around a line, this connection may be expressed by the equation of an exponential curve $y = ax^n + b$.

The fitting of the straight line through points in a logarithmic co-ordinate can be done by eye (by inspection); this line of best fit is very subjective, however, and can lead to considerable errors in determining the constants of an equation. The line may be fitted more accurately by using the rule that it must pass through a point A whose co-ordinates are:

$$x_A = \frac{\Sigma x_i}{m} ; \qquad y_A = \frac{\Sigma y_i}{m} . \tag{4.7}$$

where m is the total number of plotted points. This method is used first in this example of plotting the rating curve. From the table of measured values, it may be seen that, for $H = 0$, the discharge is 9.8, so in the exponential curve (4.5) $b = 9.8$ and

$$Q = aH^n + 9.8$$

or

$$(Q - 9.8) = aH^n,$$

after taking logarithms

$$\log (Q - 9.8) = \log a + n \log H.$$

The values of $\log (Q - 9.8)$ and $\log H$ are in *Table 9*.

Table 9

No.	$Q - 9.8$	$\log (Q - 9.8)$	H	$\log H$
1	6.1	0.7853	0.49	—0.3098
2	13.3	1.1239	0.78	—0.1079
3	20.0	1.3010	0.88	—0.0555
4	22.0	1.3424	1.03	0.0128
5	30.4	1.4829	1.20	0.0792
6	36.4	1.5611	1.28	0.1072
7	40.0	1 6021	1.34	0.1271
8	44.7	1.6503	1.36	0.1335
9	51.2	1.7093	1.51	0.1790
10	61.2	1.7868	1.55	0.1903
11	72.7	1.8615	1.76	0.2455
12	83.4	1.9212	1.86	0.2695
13	95.3	1.9791	1.92	0.2833
14	103.60	2.0154	2.10	0.3222
		Σ 22.1223		Σ 1.4764

From this table, the co-ordinates of point A can be calculated according to eq. (4.7)

$$\log H_A = \frac{1.4764}{14} = 0.1055; \qquad H_A = 1.275,$$

$$\log (Q_A - 9.8) = \frac{22.1223}{14} = 1.5802,$$

$$(Q_A - 9.8) = 38.04, \qquad Q_A = 47.84.$$

If the individual points and point A are plotted in a double logarithmic scale paper (*Fig. 4.2*), by rotation of a line around point A, the most suitable position of the line of best fit may be found. This position is indicated in *Fig. 4.2b*. The parameters of the equation can now be measur-

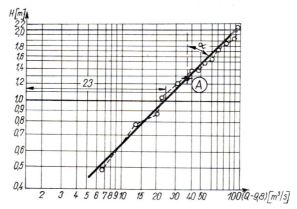

Fig. 4.2b. Stage-discharge curve (log-log paper)

ed. Parameter a will be the distance from the origin of intersection of the line with the x-axis, that is, 23. Parameter n will be the tangent of the angle between the line and the y-axis, using any rectangular triangle formed by the line with the co-ordinates. The ratio of the real distances used for calculating the tangent measured in centimetres must be trans- formed according to the scales. From *Fig. 4.2b*, it is evident that the scale of the y-axis is twice that of the x-axis.

$$\text{Ordinate } (Q - 9.8) = 3.35 \text{ cm.}$$
$$\text{Abscissa} \qquad (H) = 3.45 \text{ cm.}$$
$$\text{Hence } \tan x = \frac{3.35}{\dfrac{3.45}{2}} = 1.94 = n,$$

If parameters a and n are substituted in equation (4.6), the resulting equation is

$$\log (Q - 9.8) = \log 23 + 1.94 \log H$$

and after taking antilogarithms and adjusting, the equation of the rating curve will be

$$Q = 23H^{1.94} + 9.8.$$

Although the fitting of the curve has been made more precise with the help of the above-mentioned calculations of the average point A, the subjective factor involving human error is still considerable in this method.

In order to rule out this factor completely, a statistical method may be used, either by correlation or by the use of least squares.

(b)   Correlation and least-squares methods

The correlation method ascertains mathematically whether or not there is a relation between two or more phenomena and what its degree is.

Generally speaking, the correlation may be expressed by an arbitrary analytic curve. In most cases, however, it is advantageous and time-saving if it can be expressed by a straight line, that is, by a *linear correlation* or regression.

If there are two series of observed phenomena

$$x_1, x_2, x_3, x_4, \ldots, x_n,$$
$$y_1, y_2, y_3, y_4, \ldots, y_n,$$

(in which each x_1 corresponds to a y_1) and it may be assumed that there is a linear correlation between them (or a relation which can be transformed into a linear one, by taking logarithms, for instance), the linear regression analysis is used.

The theory of correlation and regression analysis is not reproduced here: it may be found in many reference books and manuals, in particular Chow (1964), Linsley et al. (1958), Ezekiel et al. (1959). Today, most correlation analysis is done by electronic computers and ready-made programs (software) are available in program libraries.

Only the direct and simplest practical application is therefore indicated in the following paragraphs. It is assumed that this will be of use to readers who have no access to more sophisticated tools than a simple desk calculating machine.

The linear regression of the relation between variable x and variable y (values of phenomenon A and phenomenon B) is expressed by the equations of two straight regression lines

$$y = a_1 x + b_1; \qquad x = a_2 y + b_2. \qquad (4.8)$$

The coefficients a_1, b_1, a_2, b_2 are computed from the equations

$$
\left.
\begin{aligned}
a_1 &= \frac{\Sigma(x_i - \bar{x})(y_i - \bar{y})}{\Sigma(x_i - \bar{x})^2}, \\
b_1 &= \bar{y} - a_1 \bar{x}, \\
a_2 &= \frac{\Sigma(x_i - \bar{x})(y_i - \bar{y})}{\Sigma(y_i - \bar{y})^2}, \\
b_2 &= \bar{x} - a_2 \bar{y},
\end{aligned}
\right\} \qquad (4.9)
$$

in which $\bar{x} = \dfrac{\Sigma x_i}{n}$, $\bar{y} = \dfrac{\Sigma y_i}{n}$ and n is the total number of elements in the series. a_1 and a_2 are known as coefficients of regression. The *correlation coefficient* $r = \pm \sqrt{a_1 a_2}$ can be calculated from eq. (4.9)

$$r = \frac{\Sigma(x_i - \bar{x})(y_i - \bar{y})}{\sqrt{\Sigma(x_i - x)^2 \Sigma(y_i - \bar{y})^2}}. \qquad (4.10)$$

The correlation coefficient expresses the degree of association of one phenomenon with the other and is the mathematical expression of the density of the concentration of points around the regression line.

If $a_1 = 1/a_2$, both lines are identical and merge. In such a case, the relation between both phenomena is functional. The coefficient of correlation $r = \pm\sqrt{a_1 a_2}$ is, in this case, equal to ± 1. The coefficient therefore varies between -1 and $+1$. In hydrology, the following relations between phenomena may be ascertained on the basis of the values of the coefficient of correlation:

$r = 1$ direct functional dependence,
$0.6 < r < 1$ good direct correlation,
$0 < r < 0.6$ insufficient direct correlation,
$r = 0$ no correlation,
$-0.6 < r < 0$ insufficient reciprocal correlation,
$-1 < r < -0.6$ good reciprocal correlation,
$< r = -1$ reciprocal linear functional dependence.

A direct or directly proportioned dependence means that the increase of one of the phenomena results in the increase of the other; a reciprocal or reciprocally proportional dependence means that, by the increase of one of the phenomena, the other decreases. By using r, the regression coefficients a_1 and a_2, or R_{yx} and R_{xy} of the straight regression lines, can be expressed as

$$a_1 = r \frac{\sigma_y}{\sigma_x} = R_{yx},$$

$$a_2 = r \frac{\sigma_x}{\sigma_y} = R_{xy}, \tag{4.11}$$

where σ_x, σ_y are standard deviations of the series. The straight regression lines may be now also written as

$$(y - \bar{y}) = R_{yx}(x - \bar{x})$$

and $\tag{4.12}$

$$(x - \bar{x}) = R_{xy}(y - \bar{y}).$$

A *residual* is the difference between the observed value and the value computed by using one or the other regression line (x versus y or vice-versa). Thus a standard deviation of residuals may be computed by equations:

$$\sigma_x = \sqrt{\frac{\Sigma(x_1 - \bar{x})^2}{n - 1}},$$

$$\sigma_y = \sqrt{\frac{\Sigma(y_1 - \bar{y})^2}{n - 1}},$$

and the standard deviation of distances from the regression lines is given by the equation

$$S_{yx} = \sigma_y \sqrt{1 - r^2}; \qquad S_{xy} = \sigma_x \sqrt{1 - r^2}, \tag{4.13}$$

The standard deviation (probable error) in calculating the coefficient of correlation is obtained from the equation

$$e_r = \pm 0.674 \, \frac{1 - r^2}{\sqrt{n}} \, . \tag{4.14}$$

The coefficient will thus be in the confidence interval $r \pm e_r$. In the same way that the correlation between two variables is obtained, the correlation between three or more variables can also be determined. The relevant theoretical and practical guidance can be found in the references indicated above.

The method described can obviously also be used to determine regressions other than linear, if these are transformed to a suitable linear form. This is a part of the following problem on calculation of a stage discharge curve with data used in *problem 3*.

Problem 4

There is an obvious linear correlation between the logarithms of variables Q and H. The equation of the regression line can thus be written according to eq. (4.8):

$$\log (Q - 9.8) = a_1 H + b_1 ,$$
$$\log H = a_2 \log (Q - 9.8) + b_2 .$$

To simplify, the equation is rewritten

$$\log (Q - 9.8) = y; \qquad \log H = x.$$

The calculation of values $(y - \bar{y})$, $(y - \bar{x})$, $(y - \bar{y})^2$, $(x - \bar{x})^2$ and $(x - \bar{x})(y - \bar{y})$ is included in *Table 10*.

$x = \log H$ and $y = \log (Q - 9.8)$ have already been calculated in *Table 9*.

The following values are also known from *Table 9;*

$$\bar{x} = \frac{1.4764}{14} = 0.1055; \qquad \bar{y} = \frac{22.1223}{14} = 1.5802.$$

First, the coefficient of the correlation is computed according to eq. (4.10):

$$r = \frac{\Sigma(x_i - \bar{x})(y_i - \bar{y})}{\sqrt{\Sigma(x_i - \bar{x})\Sigma(y_i - \bar{y})^2}} = \frac{0.7870}{\sqrt{1.5926 \times 0.3219}} = 0.995.$$

Table 10

No.	$y - \bar{y}$	$x - \bar{x}$	$(y - \bar{y})^2$	$(x - y)^2$	$(y - \bar{y})(x - \bar{x})$
1	—0.7949	—0.4153	0.6319	0.1725	0.3301
2	—0.4563	—0.2134	0.2082	0.0455	0.0974
3	—0.2792	—0.1610	0.0779	0.0259	0.0449
4	—0.2378	—0.0927	0.0565	0.0086	0.0220
5	—0.0973	—0.0263	0.0095	0.0007	0.0026
6	—0.0191	0.0017	0.0004	0.0000	0.0000
7	0.0219	0.0216	0.0005	0.0005	0.0005
8	0.0701	0.0280	0.0049	0.0008	0.0020
9	0.1291	0.0735	0.0167	0.0054	0.0095
10	0.2066	0.0848	0.0427	0.0072	0.0175
11	0.2813	0.1400	0.0791	0.0196	0.0394
12	0.3410	0.1640	0.1163	0.0269	0.0559
13	0.3989	0.1778	0.1591	0.0316	0.0709
14	0.4352	0.2167	0.1894	0.0469	0.0943
m	(—0.0005)	(—0.0006)	1.5931	0.3921	0.7870

Its value ($r = 0.995$) is very close to 1, the correlation ratio is very good and the points are very near the regression line.

The standard deviations of residuals are:

$$\sigma_x = \sqrt{\frac{\Sigma(x_i - \bar{x})^2}{n - 1}} = \sqrt{\frac{0.3021}{13}}$$

and

$$\sigma_y = \sqrt{\frac{\Sigma(y - \bar{y})^2}{n - 1}} = \sqrt{\frac{1.5931}{13}}$$

and their ratios

$$\frac{\sigma_y}{\sigma_x} = \sqrt{\frac{1.5921}{0.3221}} = 2.010.$$

$$\frac{\sigma_x}{\sigma_y} = \sqrt{\frac{0.3221}{1.5931}} = 0.497.$$

The regression coefficients are given by eq. (4.11):

$$R_{yx} = a_1 = r\frac{\sigma_y}{\sigma_x} = 0.995 \times 2.010 = 2.00,$$

$$R_{xy} = a_2 = r\frac{\sigma_x}{\sigma_y} = 0.995 \times 0.497 = 0.494.$$

The equation of the regression lines will therefore be, according to eq. (4.12):

$$y - 1.5801 = 2.010\ (x - 0.1055),$$

$$x - 0.1055 = 0.497(y - 1.5802);$$

and after adjustment the first equation is

$$y = 2.201x + 1.3464.$$

Since 1.3464 is the log of 22.21 and when y and x are replaced by the original variables $y - \log(Q - 9.8)$, $x = \log H$,

$$\log(Q - 9.8) = 2.01 \log H + \log 22.21$$

and the final form of the rating curve equation is

$$Q = 9.8 + 22.21H^{2.01}.$$

The standard deviation of variable y, that is of discharge $(Q - 9.8)$, can be computed from eq. (4.13):

$$S_{yx} = \sigma_y\sqrt{1 - r^2} = \sqrt{\frac{1.5931}{13}}\sqrt{1 - 0.995^2} \cong \pm0.35;$$

±0.35 is the logarithm of the deviation and so

$$S_{(Q-9.8)} = 2.2 \qquad [\text{m}^3/\text{s}].$$

In view of the values of Q (varying from 9.8 to 113.4), the standard deviation is very small, especially for large discharges.

If both equations are compared, the fitting in problem 3 is sufficiently precise for general purposes. Thus, the correlation would be used for special purposes and for when the degree of association is important, that is, where the coefficient of correlation is necessary to ascertain the existence of a dependence.

The least-squares method may also be used for accurate extension of the concentrated points. Its theory is described in detail in the references and also in books on geodetic surveying, since it is frequently used to correct surveying errors.

It is not so suitable for hydrological purposes because it is as complicated as the correlation method, but does not give the degree of association which, in the case of most calculations of hydrological analogy, is one of the most important criteria for selecting different factors in computing hydrological characteristics.

Nevertheless, the need may arise for its use, especially if it is impossible to use a linear graphical fitting or linear regression. This is the case, for instance, if a quadratic parabola is used as the fitting curve. Its equation is

$$y = a + bx + cx^2.$$

Data from problems 3 and 4 are used to illustrate this method.

Problem 5

Assume that the points of the stage discharge curve *(Table 8, Fig. 4.2a)* are best fitted by a parabola of the second degree.

According to the least-squares method, if variable y is a function of x and of parameters a, b, c, ... the values of the parameters of the best fitted curve will be best selected if the sum of the squares of the differences between the measured points and the corresponding points on the curve is a minimum. From functional analysis, it is known that this condition will be fulfilled if the partial derivatives of the differences with respect to the individual parameters will be equal to zero, that is,

$$\frac{\partial \Sigma(y - y_i)^2}{\partial a} = 0; \qquad \frac{\partial \Sigma(y - y_i)^2}{\partial b} = 0 \ldots \text{etc.}$$

From these equations the so-called *normal equations* can be derived and their solution given the parameters of the best fitted functions.

For the function $y = a + bx + cx^2$ these normal equations are:

$$\frac{\partial(y - y_i)^2}{\partial a} = \frac{\partial(a + bx + cx^2 - y_i)^2}{\partial a} = 2\Sigma(a + bx - cx^2 - y_i)x = 0,$$

$$\frac{\partial(y - y_i)^2}{\partial b} = \frac{\partial(a + bx + cx^2 - y_i)^2}{\partial b} = 2\Sigma(a + bx - cx^2 - y_i)x = 0,$$

$$\frac{\partial (y - y_i)^2}{\partial c} = \frac{\partial (a + bx + cx^2 - y_i)^2}{\partial c} = 2 \Sigma (a + bx - cx^2 - y_i) x = 0.$$

The solution of normal equations gives

$$na + b \Sigma x + c \Sigma x^2 = \Sigma y,$$
$$a \Sigma x + b \Sigma x^2 + c \Sigma x^3 = \Sigma xy,$$
$$a \Sigma x^2 + b \Sigma x^3 + c \Sigma x^4 = \Sigma x^2 y,$$

(4.15)

where n is the number of variates in the series.

If $x = H$ and $y = Q$, the following system of equations must be solved:

$$na + b \Sigma H + c \Sigma H^2 = \Sigma Q,$$
$$a \Sigma H + b \Sigma H^2 + c \Sigma H^3 = \Sigma HQ,$$
$$a \Sigma H^2 + b \Sigma H^3 + c \Sigma H^4 = \Sigma H^2 Q.$$

The values of ΣH, ΣH^2, ΣH^3, ΣH^4, ΣQ, (QH and QH^2) are calculated in *Table 11*.

The computed values of parameters a, b, c inserted in eq. (4.15) give:

$$14.0000a + 19.0600b + 28.7436c = 817.500,$$
$$19.0600a + 28.7436b + 46.5056c = 1293.659,$$
$$28.7436a + 46.5056b + 79.1289c = 2173.677.$$

From these three equations, we calculate

$$a = 17.5; \qquad b = -15.8; \qquad c = 30.4.$$

The equation of the rating curve in the form of a quadratic parabola, as derived by the least-squares method, will therefore be

$$Q = 17.5 - 15.8H + 30.4H^2.$$

We can also determine the accuracy with which the individual values of Q will be calculated, that is, the standard quadratic deviations according to eq. (4.2):

$$\sigma = \Delta Q = \pm \sqrt{\frac{\Sigma (Q - Q_v)^2}{n - 1}}.$$

Values for $(Q - Q_v)^2$ are calculated so that, for a given H, Q_v is taken from the rating curve and the difference $Q - Q_v$ is computed. Q is the

Table

1	2	3	4	5
No.	H	H^2	H^3	H^4
1	0.49	0.2401	0.1176	0.0576
2	0.78	0.6084	0.4746	0.3702
3	0.88	0.7744	0.6815	0.5997
4	1.03	1.0609	1.0927	1.1255
5	1.20	1.4400	1.7280	2.0736
6	1.28	1.6384	2.0971	2.6843
7	1.34	1.7956	2.4061	3.2242
8	1.36	1.8496	2.5155	3.4211
9	1.51	2.2801	3.4430	5.1989
10	1.55	2.4025	3.7239	5.7720
11	1.76	3.0976	5.4518	9.5952
12	1.86	3.4596	6.4349	11.9689
13	1.92	3.6864	7.0779	13.5896
14	2.10	4.4100	9.2610	19.4481
Σ	19.06	28.7436	46.5056	79.1289

measured discharge from *Table 8*. The calculation is included in *Table 11*, columns 9, 10 and 11. After substitution

$$\sigma = \Delta Q = \pm\sqrt{\frac{114.48}{13}} = \pm 2.96 \text{ m}^3/\text{s}.$$

Hence, we see that the fitting of a quadratic parabola for the rating curve is less accurate than the fitting of the exponential curve in *problem 4*.

The following equations may be used, for fitting with the least-squares method: for linear function $y = ax + b$

$$a\Sigma x^2 + b\Sigma x = \Sigma xy,$$
$$a\Sigma x + nb = \Sigma y. \tag{4.16}$$

11

6	7	8	9	10	11
Q	QH	QH^2	Computed Q_v	Residual $Q - Q_v$	$(Q - Q_v)^2$
15.9	7.791	3.818	17.1	—1.2	1.44
23.1	18.018	14.054	23.7	—0.6	0.36
29.8	26.224	23.077	27.1	2.7	7.29
31.8	32.755	33.737	33.5	—1.7	2.89
40.2	48.240	57.888	42.3	—2.1	4.41
46.2	59.136	75.694	47.1	—0.9	0.81
49.8	66.731	89.421	50.9	—1.1	1.21
54.5	74.120	100.803	52.2	2.3	5.29
61.0	92.110	139.086	63.0	—2.0	4.00
71.0	110.050	170.577	66.0	5.0	25.00
82.5	145.200	255.552	83.9	—1.4	1.96
93.2	173.352	322.435	93.3	—0.1	0.01
105.1	201.792	387.441	99.2	5.9	34.81
113.4	238.140	500.094	118.4	—5.0	25.0
817.5	1293.659	2173.677	—	—	114.48

For the exponential function $y = ax^m$, after logarithmic transformation into

$$\log y = \log a + m \log x,$$

the least-squares method equation will be

$$m \, \Sigma (\log x)^2 + \log a \, \Sigma \log x = \Sigma (\log x \log y),$$
$$m \, \Sigma (\log x) + n \log a = \Sigma (\log y). \qquad (4.17)$$

(c) Frequency distribution analysis

On *page 131*, the hydrological phenomena were characterized as random variables, continuous or discrete. The observed values constitute

Table 12

Days	Months											
	XI	XII	I	II	III	IV	V	VI	VII	VIII	IX	X
1	102	57	19	73	195	135	65	42	62	248	19	33
2	105	53	22	75	208	130	63	38	63	235	18	36
3	107	56	21	77	215	128	58	39	64	226	15	37
4	103	52	19	78	228	122	55	38	65	21	12	37
5	79	48	22	79	235	118	53	37	67	193	18	38
6	78	46	23	79	241	115	52	32	68	175	23	39
7	78	46	25	79	243	113	54	33	71	152	24	43
8	74	43	24	82	245	110	55	34	72	138	25	45
9	70	42	23	83	247	105	55	35	74	116	26	46
10	65	43	22	86	248	102	50	32	75	110	27	46
11	65	41	27	88	250	98	53	34	76	105	28	47
12	63	42	25	90	256	95	49	39	76	102	23	47
13	63	41	23	91	250	94	49	33	78	79	26	46
14	62	43	25	92	246	90	52	35	85	75	23	48
15	66	42	29	93	242	85	43	38	96	71	25	41
16	61	41	28	94	238	83	50	39	102	70	27	50
17	62	30	28	95	235	81	49	36	104	65	26	52
18	63	38	32	101	234	77	52	37	106	62	29	54
19	58	38	35	102	230	77	48	39	110	58	32	56
20	56	39	38	104	228	76	47	33	112	58	35	62
21	58	35	38	105	219	77	48	36	112	57	34	64
22	52	28	42	108	210	73	50	42	117	55	36	65
23	54	29	43	115	198	65	51	46	145	56	37	65
24	55	30	44	129	195	68	53	48	182	53	38	65
25	54	26	44	145	190	70	51	49	205	55	33	68
26	52	25	45	163	185	71	48	52	222	56	35	63
27	53	25	45	170	179	72	45	53	243	49	36	69
28	58	21	48	182	159	73	47	54	246	45	33	72
29	56	22	59		155	69	49	56	217	32	34	75
30	55	23	65		150	70	45	58	251	28	35	83
31		23	70		145		46		259	21		95

a *time series*, generally arranged in chronological order. The random character of the data in time and space is first ascertained so that each value can be treated as independent of the others. Two successive floods, of which one contributes to the peak flow of the other, constitute a time dependence; data from two adjacent rainfall stations which measure the same rainfall sample are correlated or space dependent. Thus, they cannot be included in a series for routine statistical processing. The *complete duration series* may either be processed in its original form or be transformed by selection of some particular data into a *partial duration series* or an *annual series*. A partial duration series includes only values above (or below) a certain limit, fixed for a particular purpose. An annual series is automatically constituted if a series of annual values or means is processed (annual precipitation, annual flows), or it may result from a selection of annual maxima (or minima). As will be shown later, the results of frequency analysis are not considerably influenced, as far as the extreme phenomena are concerned, by using either the partial duration or the annual series.

Thus, hydrological series will be either longer (such as daily flows) or shorter (annual floods).

When a longer series is to be processed, a frequency distribution curve may be computed and plotted as illustrated in the following problem.

Problem 6

In *Table 12*, 365 values of daily flows in one year are tabulated chronologically; they form a complete duration series. Compute the frequency distribution curve of daily flows and the flow duration curve.

The variable in this case is a continuous one: flows in m^3/s. The amplitude of the series is the difference between the maximum and minimum variate, e.g. 12 m^3/s and 259 m^3/s respectively (*see Table 12*). This amplitude may be divided into intervals and the values of the variable distributed in these intervals. The number of occurrences in each of the intervals is called the *frequency n*. Divided by the total number of variates N in the series, it is the *relative frequency f* in per cent. These are computed in columns 2 and 3 of *Table 13*. If plotted in arithmetic coordinates, a frequency distribution bar graph (histogram) is obtained (*see Fig. 4.3*) .

Table 13

1	2	3	4	5	6
	Frequency *n*		Cumulative frequency (duration)		
Intervals [m₃/s]	*n* (daily flows)	$\dfrac{n}{N}$ [per cent]	Σn (daily flows)	$\dfrac{\Sigma n}{N}$ [per cent]	Notes
259—240	16	4.4	16	4.4	
239—220	10	2.7	26	7.1	
219—200	6	1.6	32	8.7	
199—180	8	2.2	40	10.9	Maximum 31.7
179—160	4	1.1	44	12.0	259 m³/s
159—140	7	1.9	51	13.9	Minimum 4.9
139—120	6	1.6	58	15.5	12 m³/s
119—100	27	7.4	84	22.9	
99—80	21	5.8	105	28.7	
79—60	66	18.1	171	46.8	
59—40	98	26.9	268	73.6	
39—20	89	24.4	358	98.1	
19—0	7	1.9	365	100.0	
Totals	365	100%			

For the sake of simplicity, equal intervals were chosen. If they were different, each value of frequency would have to be divided by the size of the interval b. This ratio is called the frequency density $\bar{h} = \dfrac{n}{b}$ or $\bar{h}$ per cent $= \dfrac{f}{b}$. It is obvious that, using equal intervals, the frequency is always proportional to frequency density.

If the variate is discrete, no intervals are necessary, since a frequency exists for each possible value of the variate.

In hydrology (and in frequency analysis in general), it is usually necessary to ascertain how many occurrences of the variate have a higher (or lower) value than a certain given value. Thus, for example, in *problem 6* it would be of interest to know how many daily flows were higher (or lower) than a flow which corresponds to a bankful flow in the channel; in other words, for how many days of the year the river will overflow the banks and flow in the flood plain. Such information may be necessary for the design of a channel control cross-section.

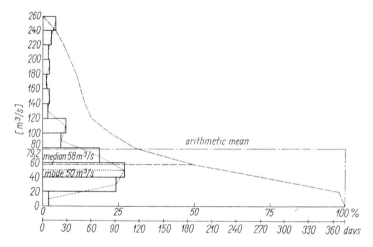

Fig. 4.3. Distribution graph and duration curve

This information may be ascertained from the *cumulative frequency* or its relative value, which is obtained by summation of the frequencies of occurrence in each interval (*see columns 4 and 5 of Table 13*). Plotted in a graph (*Fig. 4.3*), a cumulative frequency graph or a *duration curve* is obtained. If the period of observation is one year, as in *problem 6*, and each variate represents a daily value, the duration curve indicates for how many days during the year the flow has exceeded (or not attained) any given value in the variation amplitude.

The construction of the duration curve from the histogram may be also a graphical summation of the abscissae of the distribution graph.

The frequency distribution graph indicates several other characteristics of the series. The value of the variate having the highest frequency, or the maximum abscissa of the frequency distribution curve, is the *mode*. The value of the variate having a 50 per cent cumulative frequency, in other words, half of the variates being below it and the other half above, is the *median* of the series. Finally, the arithmetic mean may also be ascertained from both the frequency distribution graph and the duration curve. In the first case, the abscissa passing through the centre of gravity of the graph indicates on the axis of the ordinates the value of the arithmetic mean. In the second case, it is indicated by the abscissa which halves the area of the duration curve (*see Fig. 4.4*).

To sum up, *Fig. 4.3* and *Table 13* are the solution of *problem 6*.

The construction of frequency distribution graphs and duration curves is mostly used for historical data such as water stages and discharges. However, a theoretical approach to this problem permits further development of the hydrological characteristics, in particular with respect to their prediction in the future. In this connection, the concept of theoretical frequency distribution and probability is examined below.

Theoretical frequency distribution

In the histogram, the density of frequency in each interval is an average. In fact, each interval may be divided into an arbitrary number of smaller intervals, each of them having a different frequency density, and the frequency density in the original interval is an average value of all of them. To each value of variate x will thus correspond a cumulative frequency $F(x)$. In an arbitrary interval from x to $x + b$, this cumulative frequency will be $F(x + b) - F(x)$ and the average frequency density in the interval will be

$$\bar{h} = \frac{F(x + b) - F(x)}{b}.$$

For a continuous variable, if the interval b becomes infinitesimally small while the number of variates remains finite, there will be in some intervals a number of variates, while in others there will be none. The cumulative frequency in these last intervals will be equal to 0.

Thus, the cumulative frequency for a series with a finite number of variates N increases discontinuously (in jumps), at least by $\dfrac{1}{N}$ in each point corresponding to the value of every variate of the series.

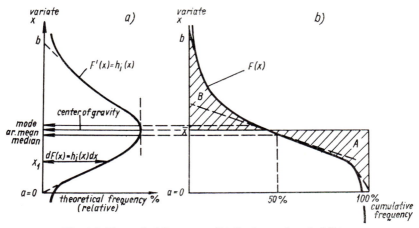

Fig. 4.4. Theoretical frequency distribution and probability curves

If the number of variates n increases to infinity (as in a theoretical mathematical series) and the interval b decreases infinitesimally, the cumulative frequency function $F(x)$ will have a derivative $F'(x)$ related to $F(x)$ for $n \to \infty$ by the equation

$$\lim_{b \to 0} \frac{F(x + b) - F(x)}{b} = F'(x) = h_i(x)$$

where $h_i(x)$ is a theoretical density of frequency and the differential $dF(x) = h_i(x) \cdot dx$ is the theoretical frequency in the interval dx.

The curve expressing the theoretical frequency distribution will then be continuous, and will have a form similar to the histogram (*see Fig. 4.4a*).

By integrating the function of theoretical frequency distribution, the *theoretical cumulative frequency function* $F(x)$ will be obtained which can be represented by a continuous curve (*see Fig. 4.4b*). Thus

$$F(x) = \int F'(x)\, dx.$$

In general, all the variates are contained in the interval $-\infty$ to $+\infty$. The value of the above integral will therefore be always 100 per cent or 1 (in absolute value). In hydrology, however, the phenomenon cannot have a negative value and often cannot be superior to a maximum value a. Thus the integral may be written

$$F(x) = \int_0^a F'(x)\,dx = 1.$$

The theoretical curve of frequency distribution has similar parameters to the histogram. *Arithmetical mean, mode* and *median* may be ascertained for it (*see Fig. 4.4a* and *b*).

Probability

The definition of probability may be derived either from the statistical approach to frequency distribution or from the basic logical probability approach.

In the first case, the probability is defined as the cumulative frequency of a variate if the number of variates in the series increases infinitely. Thus, the probability is identical with the theoretical cumulative frequency as derived above:

$$P(x) = \int F'(x)\,dx.$$

In the second approach, probability is the measure of an objective possibility of occurrence of a random phenomenon. If n is the number of possibilities of occurrence of this phenomenon (or the number of variates in the series) and m the number of occurrences of this phenomenon (for a continuous variable it is understood that the value of the variate was attained in m number of cases), the probability is the ratio of actual occurrences to the total number of possible occurrences

$$P(x) = \frac{m}{n}.$$

The probability thus defined varies from 0 to 1. Indeed, if the value of the phenomenon has never occurred, its probability is $\frac{0}{n}$ or 0. If the value was attained in every occurrence of the phenomenon, its number of occurrences is $m = n$ and its probability is $\dfrac{m}{n} = \dfrac{n}{n} = 1$.

This definition and computation of probability is strictly applicable to closed series only. Time series in hydrology are seldom closed, since it is not possible to ascertain all values of occurrences of the phenomenon in the past.

On the other hand, this definition of probability, keeping in view its derivation from theoretical cumulative frequency, provides a tool to ascertain duration curves (or cumulative frequency curves; also probability curves) for shorter hydrological time series directly, without passing through the frequency distribution graph (histogram) or curve. This is illustrated in *problem 7*.

Problem 7

In *Table 14* is annual precipitation hP [mm] at station T for the period 1934 — 1954. Ascertain the probability (cumulative frequency)

Table 14

Year	1934	1935	1936	1937	1938	1939	1940	1941	1942	1943	1944
hP [mm]	460	443	542	609	568	741	637	724	433	324	563

Year	1945	1946	1947	1948	1949	1950	1951	1952	1953	1954	
hP [mm]	645	451	515	474	376	605	426	530	341	666	

curve of this series. The series includes 21 variates ranging from 324 mm to 741 mm. The approach used in problem 6 is not possible here, owing to the low number of variates.

If the series is arranged in descending order of magnitude (*Table 15, column 3*) it is evident that the highest value (741 mm) was attained and

Table 15

1	2	3	4	5	6	7	8
No.	Year	hP [mm]	$P = \dfrac{m-0.3}{n+0.4} 100$	$k = \dfrac{x}{\bar{x}} i$	$(k-1)$	$(k-1)^2$	$(k-1)^3$
1	1939	741	3.27	1.40	0.40	0.1600	0.0640
2	1941	724	7.94	1.37	0.37	0.1369	0.0506
3	1954	666	12.6	1.26	0.26	0.0676	0.0176
4	1945	645	17.3	1.22	0.22	0.0484	0.0106
5	1940	637	22.4	1.21	0.21	0.0441	0.0093
6	1937	609	26.6	1.16	0.16	0.0256	0.0041
7	1950	605	31.3	1.15	0.15	0.0225	0.0034
8	1938	568	36.0	1.08	0.08	0.0064	0.0005
9	1944	563	40.6	1.07	0.07	0.0049	0.0003
10	1946	551	45.3	1.05	0.05	0.0025	0.0001
11	1936	542	50.0	1.03	0.03	0.0009	0.0000
12	1952	530	54.6	1.01	0.01	0.0001	0.0000
13	1948	474	59.4	0.90	—0.10	0.0100	—0.0010
14	1934	460	64.0	0.87	—0.13	0.0169	—0.0022
15	1935	443	68.7	0.84	—0.16	0.0256	—0.0041
16	1942	433	73.4	0.82	—0.18	0.0324	—0.0058
17	1951	426	78.0	0.81	—0.19	0.0361	—0.0069
18	1947	414	82.7	0.78	—0.22	0.0484	—0.0106
19	1949	376	87.4	0.71	—0.29	0.0841	—0.0243
20	1953	341	92.1	0.65	—0.35	0.1225	—0.0429
21	1943	324	96.7	0.61	—0.39	0.1521	—0.0593
Σ		11 072			0.00	1.0080	0.0034

thus occurred only once, and its $m = 1$. The total number of occurrences was 21, consequently $n = 21$. Its probability is

$$p = \frac{m}{n} = \frac{1}{21} = 0.048. \qquad (4.18)$$

The same procedure applied to the second highest value (724 mm), which was attained (occurred or was exceeded) twice, hence $m = 2$, its probability

$$p = \frac{2}{21} \cong 0.095.$$

The lowest, twenty-one, value's probability will be

$$p = \frac{21}{21} = 1.$$

A probability 1 indicates that such a value of the phenomenon is *certain* to be *always* attained. This, while correct for a closed series, is evidently erroneous for a series such as the one indicated in this problem. Indeed, a lower annual precipitation may have occurred in the unrecorded past and certainly may occur in the future. If applied, for example, to the lowest value of the variate in *problem 6*, the probability p (since the series is closed) is

$$p = \frac{365}{365} = 1.$$

The minimum value of the discharge during this year was indeed 12 m^3/s and no lower value occurred that year.

While, for the derivation of the duration curve in *problem 6*, the equation (4.18) $p = \frac{m}{n}$ may be used with justification, this equation will be not suitable for *problem 7*.

Theoretical works in frequency analysis (Chow, 1964; Dzubák, 1960) have proposed changing the above equation so as to serve purposes of frequency analysis of relatively short time series, representing a relatively limited sample of the whole population.

Such formulae are

$$p = \frac{m}{n + 1} \qquad \text{(Weibull)} \qquad\qquad (4.19)$$

and

$$p = \frac{m - 0.3}{n + 0.4} \qquad \text{(Chegodayev)} \qquad\qquad (4.20)$$

Both are suitable for processing annual and extreme-value time series. Chegodayev's formula is widely used in Czechoslovakia and prescribed as a standard for flood peak computations in the USSR. It is used for the solution of *problem 7*. In *Table 15, column 4*, the values of probability of each variate are computed and the values are plotted in *Fig. 4.5* on probability paper, of which the use and construction is explained below.

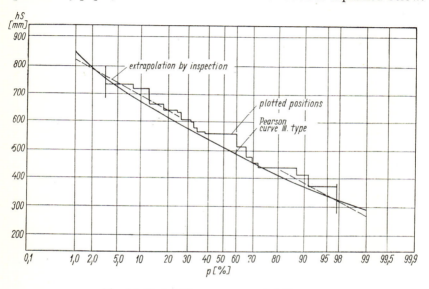

Fig. 4.5. Probability curve on probability paper

Recurrence interval

The recurrence interval T (return period, periodicity) of a hydrological phenomenon is the *average* interval of time within which the value (magnitude) of the event will be equalled or exceeded *once*.

Thus, if $T = 10$ years, the corresponding value of the phenomenon (flood peak, 1 h rainfall depth, etc.) will occur *on the average* once every 10 years. It is most important to retain the word 'average', since it indicates that the recurrence of the event (in this example) is not necessarily every 10 years in the chronological sense, but that the event occurs 10 times in 100 years or, on average, once every 10 years.

If the phenomenon is exceeded or equalled on average once every T years, its probability p is

$$p = \frac{1}{T} \quad \text{or} \quad T = \frac{1}{p} \tag{4.21}$$

and, reciprocally, since the probability of *not* being equalled or attained is $1 - p$, for values of the variate below the given value

$$T = \frac{1}{1 - p} \, . \tag{4.22}$$

Since the recurrence interval is usually given in years or fractions of a year, the above relation of probability and recurrence interval applies to annual series. The partial duration series recurrence interval T_E is related to the annual series recurrence interval T by the equation

$$T_E = \frac{1}{\ln T - \ln (T - 1)} \, .$$

Nevertheless, the differences between the two values, although depending on the series, are relatively significant only for recurrences below 5 years $(1 - 5 \text{ years})$ and for fractions of a year, as illustrated by the following table for conversion of recurrence intervals derived from both types of series.

Partial series		Annual series
T_E of 0.50 year	=	T of 1.16 years,
T_E of 1.00 year	=	T of 1.58 years,
T_E of 1.45 year	=	T of 2.00 years,
T_E of 2.00 year	=	T of 2.54 years,
T_E of 5.00 year	=	T of 5.52 years,
T_E of 10.00 year	=	T of 10.50 years.

The recurrence interval of the highest value of the annual series in *problem 7* will thus be (values of p from *Table 15, col. 4*):

$$T = \frac{1}{0.0327} = 30.6 \text{ years.}$$

The lowest value of the series will have a recurrence interval

$$T = \frac{1}{1 - 0.967} = \frac{1}{0.033} = 30.3 \text{ years.}$$

From *column 4* of *Table 15* it appears that the *median* value variate is the eleventh, $hP = 552$ mm. Its return period $T = \dfrac{1}{0.5} = 2$ years. The value is exceeded or not attained every second year on average.

Prediction probability

The above procedures of frequency analysis have the practical aspect of prediction.

The term prediction is used here in the sense that a probability of occurrence is derived for design purposes in the future; however, the exact (real) time of occurrence of this value is not given. The prediction in real time, indicating the magnitude of a hydrological event in a near (or distant) future at a given chronologically ascertained moment, is the subject of hydrological *forecasting*.

The frequency distributions ascertained by the use of histograms or probability-plotting formulae cover a prediction period not much longer than the period of observation in the past. Thus, if the probable value of an event with a longer return period is to be computed, extrapolation of the prediction period and thus of the probability curves is necessary. This may be achieved either by graphical extrapolation, mainly on a special graph paper — *probability paper* — or by substituting the observed frequency distribution by a *theoretical frequency distribution* (*probability*) *curve* fitted to the observed series either graphically or by computation of its parameters from the observed series.

Probability paper

The object of using this paper is to linearize the frequency distribution, mainly so that a theoretical probability curve appears on the paper as a straight line. On the ordinate of such paper the variate is generally plotted in an arithmetical or logarithmic scale; the probability p or recurrence interval T is indicated on the abscissa. There are many such papers (for Gauss — Laplace distribution, for Gumbel distribution, etc.). Two such papers are attached in Appendix 1, one for linearizing the binomial distribution (Pearson Type III), the second for Gumbel distribution. In *Fig. 4.5* the probability curve computed in *problem 7* is extrapolated, by a straight line fitted by inspection, to a return period of 100 years.

Such extrapolation is most subjective, and it may induce a considerable error of sampling. It should not be used without adequate interpretation and possible hydrometric or deterministic confirmation of the extrapolated value. This applies in particular to flood peak flows.

Theoretical probability curves and their mathematical fitting

This fitting is usually done by the computation of the parameters of the theoretical curve from observed data (*method of moments*) or by the use of the *frequency factor*. It is beyond the scope of this book to describe the theoretical approach to the selection and discussion of the suitability of different theoretical probability curves, of which a great number exists. Three such curves will be used in this book:

the Pearson Type III distribution,
the Gumbel extreme value distribution,
the Goodrich distribution (Alexeyev's version).

The warning voiced on the use of probability paper applies equally to the extrapolation by theoretical curves. The application of a safety coefficient will be indicated when appropriate.

It should never be overlooked that all theoretical curves imply the existence of an infinite sample (series of variate), a condition which is never attained in practical hydrological computations. The law of large numbers is always to be taken into consideration. According to this law, the probability that the frequency distribution is closely approximated by the theoretical probability sharply decreases with a decrease of the number of variates (particularly below 30) and equals 1 (hence, is certain) with an infinitely large number of variates.

The use of the method of moments in mathematical curve fitting will be illustrated in *problem 8*.

Problem 8

Extrapolate the probability curve from problem 7 to probabilities 0.01 (1 per cent) and 0.99 (99 per cent), e.g., of a return period of 100 years, for the events respectively exceeded and not attained on the average once every 100 years. The theoretical curve selected for the solution is the Pearson Type III distribution. This is an asymmetrical (skew) distribution, which may become a symmetrical (Gauss – Laplace) distribution if the parameter of skewness (coefficient of asymmetry) is zero.

Asymmetry of a distribution indicates that the deviations of the variate from the mean are less frequent but larger on one side of the mean than on the other.

This distribution is generally recognized as suitable for annual non-extreme series and, in the USSR, it is also used for extreme flood frequency analysis as a standard method.

Its parameters calculated from the observed series are:

(a) arithmetic average

$$\bar{x} = \frac{\Sigma x_i}{n}, \tag{4.1}$$

(b) coefficient of variation.

$$C_v = \frac{1}{\bar{x}} \sqrt{\frac{\Sigma(x_i - \bar{x})^2}{n - 1}} = \frac{\sigma}{\bar{x}}. \tag{4.23}$$

It is the ratio of the standard deviation and the arithmetic average. It is a non-dimensional parameter which makes possible a comparison of the dispersion of individual series. If x_i is replaced by its module coefficient $k_i = \frac{x_i}{\bar{x}}$, then

$$C_v = \sqrt{\frac{\Sigma(k_i - 1)^2}{n - 1}} \tag{4.24}$$

and the coefficient of skewness (asymmetry)

$$C_s = \frac{\Sigma(k_1 - 1)^3}{(n-1)\,C_v^3} \tag{4.25}$$

indicates the degree of skewness of the distribution.

The *Pearson curve* is truncated on one side of the axis of magnitude of the variate, which means that below a certain value of the variate the probability is zero, but it is infinite, converging asymptotically to the axis of the magnitude of the variate. This means that even values of

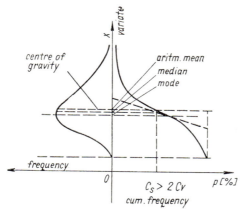

Fig. 4.6. Pearson's Type III distribution

infinitely large (or small) phenomena have a certain probability of occurrence, even though this may be infinitely small. This last characteristic of the curve is not such a handicap in its practical use as the fact that the origin of the curve can also be in the negative part of the axis of the magnitude of the variate. Thus, even the negative values of a phenomenon may have a certain probability of occurrence. This evidently does not correspond with the physical interpretation of the extrapolation because hydrological phenomena cannot have a negative value.

The position of the origin of the curve depends on the ratios of C_v and C_s.

If $C_s \geqq 2C_v$, the origin of the curve is at the origin of the co-ordinates or in the positive part of the co-ordinates.

The accuracy of the calculation of C_v is computed by the equation for error E_c of C_v

$$E_{c_v} = \pm \frac{0.674 C_v}{\sqrt{2n}} \sqrt{1 + 2C_v^2}. \qquad (4.26)$$

C_v is within the confidence interval if the series includes a minimum of 20 to 30 variates. With longer series the error of computation of the parameters of the theoretical distribution curves decreases.

The accuracy of the calculation of C_s is much less than for C_v and, for a series of 100 figures, the probable error of C_v is still ± 25 per cent. For this reason, and also in order to situate the origin of the curve at the origin of the coordinates, C_s is often assumed (instead of being calculated) as

$$C_s = 2C_v.$$

If the Type III Pearson curve is used for extrapolating extremes (flood peak discharges, for instance), larger ratios may be assumed – up to $C_s = 4C_v$ (see also page 243).

After computations of the three parameters above, the co-ordinates of the analytic curve are looked up in the table of the Pearson Type III (or other) functions. Appendix 4 includes this table for any C_s; appendix 5 gives the co-ordinates of the curve for the ratio $C_s = 2C_v$.

For theoretical derivation of these procedures and more details, the reader is directed to Nash, J. E.: Applied Flood Hydrology in 'River Flood Hydrology' (1969) and V. T. Chow: Handbook of Applied Hydrology, Section 8 (1969).

Calculations may be tabulated as in the solution of *problem 8*. The step-by-step procedure is indicated below.

(a) The variates of the series are arranged in descending order of magnitude (see columns 2 and 3 of Table 15).

(b) The arithmetic average is computed according to eq. (4.1), Σx_i being the sum of column 3,

$$\bar{x} = \frac{11\,072}{21} = 527.$$

(c) Calculate the reduced variate k_i according to the equation

$$k_i = \frac{x_i}{\bar{x}} \text{ (column 5)}.$$

(d) Calculate the values $k_i - 1$ (column 6). Their sum verifies the accuracy of the calculation, since it is to approximate 0.

(e) Calculate values $(k_i - 1)^2$ (column 7), and $(k_i - 1)^3$ (column 8) and their sums; $\Sigma(k_i - 1)^2 = 1.0480$; $\Sigma(k_i - 1)^3 = 0.0034$.

(f) Enter these figures into eq. (4.24) and (4.25):

$$C_v = \sqrt{\frac{1.0480}{20}} = 0.23,$$

$$C_s = \frac{0.0034}{20 \times 0.1112} = 0.0139.$$

Since $C_s < 2C_v$, considering the error in calculation of C_s, $C_s = 2C_v = 0.46$ is assumed.

Probable error of C_v, E_{C_v} is computed from eq. (4.26):

$$E_{C_v} = \frac{0.674 \times 0.23}{\sqrt{42}} \sqrt{1 + 2 \times 0.23^2} = \pm 0.025.$$

(g) The co-ordinates Φ of the Pearson Type III curve for the given C_v and C_s may be found in the table in appendix 4 in the following form:

$$\Phi_{p\%} = \frac{k_{p\%} - 1}{C_v}, \tag{4.27}$$

where $\Phi_{p\%}$ is the function of the reduced variate value for probability,
p per cent and for $\bar{x} = 1$, $C_v = 1$,
$k_{p\%} = $ the reduced variate calculated by the equation:

$$k_{p\%} = \Phi_{p\%}C_v + 1.$$

The computations are made according to eq. (4.27) in *Table 16*. If the table of $k_{p\%}$ in appendix 5 is used instead of appendix 4, the value $k_{p\%}$ is obtained directly (the figures in *the fourth line of Table 16*).

Line 5 of Table 16 gives the values of the Pearson Type III frequency distribution curve from $p = 1$ per cent to $p = 99$ per cent, fitted to annual precipitation values (hS) distribution.

The probabilities of 1 per cent and 99 per cent (recurrence interval 100 years) have been thus extrapolated by the use of a theoretical probability curve.

Table 16

Probability / Frequency factor		p [per cent]							
		1	10	30	50	70	80	95	99
2	Φ_p (per cent)	2.7	1.325	0.47	−0.07	−0.575	−0.855	−1.53	−2.06
3	$\Phi_p C_v$	0.621	0.305	0.108	−0.016	−0.132	−0.196	−0.352	−0.474
4	$(\Phi_p C_v - 1) = k_p$	1.621	1.305	1.108	0.984	0.868	0.804	0.648	0.526
5	$hP_p = \bar{x}k_p$ [mm]	854	687	583	518	457	424	341	277

Calculated for $C_s = 0.46$.

The probability curve is plotted from computed points on probability paper in *Fig. 4.5*. The curve on this paper is almost a straight line, but the ends deviate. This shows the subjective and relatively inaccurate graphic interpolation made by eye fitting.

The generally good fit between the empirical points (plotting positions) and the theoretical curve confirm the approximate choice of $C_s = 2C_v$.

If the empirical points and theoretical curve differ considerably, another ratio may be chosen between C_s and C_v ($C_s = 3C_v$, for instance) and the calculations in *Table 16* should be repeated.

Several hydrologists shorten the above procedure of computation and make use of some characteristic points of the theoretical curve for determining the parameters of the curve (method of quantils).

In his work on grapho-analytical methods (Trudyi GGI, No. 73, Leningrad, Gidrometeoizdat 1960), G. A. Alekseyev has introduced such a method for calculating the parameters of the Type III Pearson curve.

According to this method, the empirical plotting positions of the variates are calculated and plotted on probability paper. The equation

$$p = \frac{m - 0.3}{n + 0.4} \, 100 \text{ per cent is recommended for this computation.}$$

A line is fitted to the points, from which the values of the variable for $p = 5$, 50 and 95 per cent are obtained (X_5, X_{50}, X_{95}). The skewness S

is calculated from equation:

$$S = \frac{X_5 + X_{95} - 2X_{50}}{X_5 - X_{95}}. \tag{4.28}$$

In a table calculated by Alekseyev and included in appendix 3, the corresponding values of the skewness coefficient C_s and the reduced variate co-ordinates of the Pearson curve (Φ_5, Φ_{50} and Φ_{95}) may be found for a given S. The standard deviation σ, the variation coefficient C and the arithmetic average may now be calculated, by equations

$$\sigma = \frac{X_5 - X_{95}}{\Phi_5 - \Phi_{95}}, \tag{4.29}$$

$$\bar{X} = X_{50} - \sigma\Phi_{50} \tag{4.30}$$

$$C_v = \frac{\sigma}{\bar{X}}. \tag{4.31}$$

Thus, all the parameters of the probability curve are known. Any particular variate of the theoretical curve may be computed by using the tables in appendix 4 or 5, as above.

The accuracy of this method may be increased by using five, instead of three, plotted positions ($X_5, X_{10}, X_{50}, X_{90}, X_{95}$). The procedure is the same, except that eq. (4.28) and (4.29) will be:

$$S = \frac{X_5 + X_{10} + X_{90} + X_{95} - 4X_{50}}{X_5 + X_{10} - X_{90} - X_{95}}, \tag{4.32}$$

and

$$\sigma = \frac{X_5 + X_{10} - X_{90} - X_{95}}{\Phi_5 + \Phi_{10} - \Phi_{90} - \Phi_{95}}. \tag{4.33}$$

Thus, the parameter of the theoretical curve may be determined without the cumbersome calculations of the arithmetic average, k, $(k - 1)$, $(k - 1)^2$, $(k - 1)^3$ and their sums. This short-cut is therefore particularly suitable for long series.

Problem 9

Calculate the parameters of the curve as in *problem 8* using Alekseyev's short method of three points. The plotting positions were computed in *problem 7* and, in *Fig. 4.5*, the frequency distribution curve was first plotted by inspection. The values of X_5, X_{50} and X_{95} from *Fig. 4.5* are:

$$hP_{5\%} = 745 \text{ mm}, \qquad hP_{50\%} = 520 \text{ mm}, \qquad hP_{95\%} = 335 \text{ mm}.$$

These values are entered in eq. (4.28):

$$S = \frac{745 + 335 - 2 \times 520}{745 - 335} = \frac{1080 - 1040}{410} \cong 0.10.$$

From the Table in appendix 3 for $S = 0.10$, the values of $C_s = 0.4$, $\Phi_5 = 1.75$, $\Phi_{50} = -0.07$, $\Phi_{95} = -1.52$. These, entered in eq. (4.29), give:

$$\sigma = \frac{745 - 335}{1.75 + 1.52} = \frac{410}{3.27} = 125 \text{ mm}$$

and then from eq. (4.30):

$$\bar{x} = 520 + 125 \times 0.07 = 520 + 9 = 529$$

and from eq. (4.31)

$$C_v = \frac{125}{529} \cong 0.235.$$

Thus, the calculated parameters of these series are: long-term average precipitation (arithmetic average) $MhP = 529$ mm,

$$C_v = 0.235, \qquad C_s \cong 0.4.$$

These are basically the same as the parameters calculated in *problem 8*. The calculation of values of precipitation for different p will be the same as in *problem 8*.

Fitting of other distributions (Gumbel and Goodrich) is described later (*see pages 187 and 244*).

To conclude this section, attention should be drawn to the dangers of an uncritical use of mathematical statistical methods, particularly of the theoretical frequency distribution curves. It has been demonstrated that these curves are based on an infinitely large sample. By extrapolating relatively short series of observations, serious errors may be introduced, particularly if physical conditions of the watershed have changed and the natural tendencies of the processed hydrological phenomena are not taken into consideration.

The processing of precipitation data in *problem 8* may be taken as an example. In calculating the annual precipitation of 100-years recurrence interval, the period 1934 to 1955 was used. A precipitation was predicted which would occur on the average once every hundred years. However, it is known that climatic conditions on the earth evolve in long-term cycles. Dry and wet seasons alternate, as well as warm and cold periods. Precipitations which, from the short-term standpoint, may be considered as random, have, in the long term, a rising or falling tendency. Hence, if the 1934 — 1935 period is in the dry cycle, the statistically computed results will be lower than if they had been obtained from a wet cycle period. Thus a purely random character may not be assumed for them. In addition, for runoff data, human activities are of major significance. Man can completely change the physical conditions of the basin (by cutting down the forest, for instance) and thereby artificially influence the natural tendency of surface or groundwater flow.

A careful verification is recommended of the results obtained from processing by mathematical statistics, by comparing them with results obtained by deterministic methods, using factors of the hydrological cycle.

4.4   Basic hydrographic parameters of basins and streams

The area of the basin and the length of the stream are the most important of these parameters.

Problem 10

In *Fig. 4.7*, on a map of 1 : 25 000, is the stream 'J'. Ascertain the area of its basin and its length from the site of the gauging cross-section, indicated on the map by a triangle.

Assuming the elementary physical law of surface runoff, that water flows perpendicularly to contour lines, the watershed divide will be an orthogonal trajectory to the contour lines originating at the indicated site on the stream and leading to the highest elevation in the basin. If it is difficult to determine the direction of the highest elevation toward which the watershed divide runs, the drawing may start again on the other direction of the cross-section located on the stream. The watershed divide line thus drafted must be thoroughly checked to avoid mistakes leading to overlap of adjacent basins.

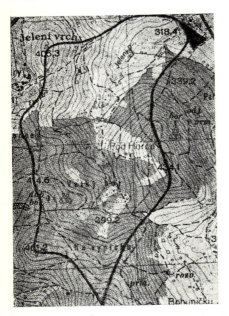

Fig. 4.7. Area of catchment

The measuring of the area of the basin on the map is done mainly by a polar planimeter. A string planimeter is also used but less frequently. Transparent graph paper may be used when these instruments are not available.

When measuring with a polar planimeter, it is necessary to ascertain the unit of the vernier. If the planimeter is not set to a map scale according to the instructions attached, this unit can be obtained very easily by measuring a square of $10 \, \text{cm}^2$ on graph paper. On a 1 : 25 000 map, this is an area of 6.25 km², on a 1 : 50 000 scale map it represents 25 km².

If b is the number of vernier units obtained by subtracting the initial vernier reading from the final one, and if n is the number of contouring of the watershed divide line, the vernier unit a will be for a 1 : 25 000 map:

$$a = \frac{n \times 6.25}{b} \quad [\text{km}^2]$$

and for a 1 : 50 000 map:

$$a = \frac{n \times 25}{b} \quad [\text{km}^2]$$

At least double or triple circuits of the divide line by the planimeter is recommended to eliminate random errors.

Planimeter readings for the basin in *Fig. 4.7* are given in *Table 17*. The planimeter vernier unit is 4.0 km² and the average vernier reading 0.433. Hence, the area of the basin is $4.0 \times 0.433 = 1.732$ km².

Table 17

Reading	Contouring of the area			
	1	2	3	average
Initial	0.000	0.422	1.300	
Final	0.422	0.864	1.735	
Difference	0.422	0.442	0.435	0.433

In measuring the length of a stream from a map, it must be kept in mind that streams are simplified when drawn on maps. Since the channel of the stream very often changes its bank line, the length of a stream can be accurately determined only by surveying in the field the streamline or the centre-line of the stream.

In the United Kingdom and US, the 'smoothed' map length is considered as adequate and measured with a wheel-type map measurer.

In the USSR and other eastern European countries, the length of a stream read from a map is multiplied by a tortuosity coefficient $1.01 - 1.10$, exceptionally 1.25.

The measurement may also be made with dividers, using 1 to 2 mm distances. The number of distances along the whole length of the stream is noted and their sum is multiplied by the above coefficient, according to the map scale.

In problem 10, a 2 mm distance on the map (50 m in nature) was chosen. A total of 45 distances were measured to the end of stream. Another 20 distances were measured in the thalweg and above the head of the stream until the divide line. The tortuosity coefficient was assumed as 1.04. The length of the stream $L = 1.04 \times 45 \times 50 = 2440$ m. The length of the whole basin along the valley is $L_1 = 1.04 \times 65 \times 50 =$ = 3380 m.

Other basin parameters are: the average width of the basin, a coefficient of the basin's shape, average slope of the basin, average and partial slopes of the main stream, the forestation coefficient of the basin and the basin's average elevation above sea-level.

The average width of the basin may be computed as the ratio of the area of basins and the length of the thalweg

$$\bar{b} = \frac{A}{L}, \tag{4.34}$$

where A = the area of the basin [km²],

L = length of the thalweg of the stream (or of the valley L_1, if the stream does not reach the watershed divide) [km].

For the basin data in *problem 10*

$$\bar{b} = \frac{A}{L_1} = \frac{1.732}{3.33} = 0.512 \text{ km}.$$

A *coefficient of the shape* of the basin d may be selected as defined by the following equation

$$d = \frac{L^2}{A}. \tag{4.35}$$

For *problem 10*, $d = \dfrac{L^2}{A} = \dfrac{3.38^2}{1.732} = 6.6.$

There are, however, several other coefficients which may express the shape of the basin.

The average slope of the basin is obtained by the equation:

$$I = \frac{h(0.5l_1 + l_2 + l_3 + \ldots + l_{n-1} + 0.5l_n)}{A}, \tag{4.36}$$

where I = the average slope of the basin in absolute units,

h = elevation difference between contour lines [km],

l_1 and l_n = respective lengths of the lowest and highest contour line in the basin [km],

$l_2 - l_{n-1}$ = length of the other contour lines in the basin [km],

A = area of basin [km²].

An approximation of this slope may be computed by the equation:

$$I = \frac{h_{max} - h_{min}}{\sqrt{A}}, \tag{4.37}$$

where h_{max} and h_{min} are the highest and lowest points in the basin, A as above.

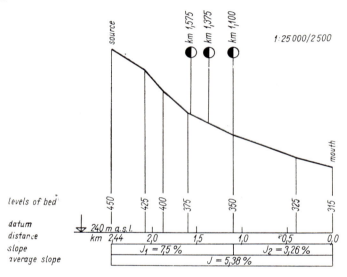

Fig. 4.8. Longitudinal section of stream

For the basin in *problem 10* equation (4.36) will yield

$$I = \frac{20(0.5 \times 0.15 + 1.075 + 1.50 + 2.15 + 2.4 + 1.3 + 0.75 + 0.3 \times 0.5 \times 0.05)}{1.732} =$$

$$= \frac{0.020 \times 9.572}{1.732} = 0.11 = 11 \text{ per cent}$$

and from eq. (4.37)

$$I = \frac{0.048\ 12 - 0.3150}{1.732} = 0.127 = 12.7 \text{ per cent.}$$

The difference in the average slope computed by both equations is not considerable in this example.

The average and partial slopes of the stream are computed from the longitudinal cross-section (*see Fig. 4.8*). It may be plotted from data read from the map. The average slope of the stream from *problem 10* is 5.38%.

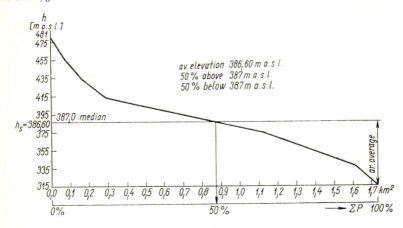

Fig. 4.9. Hypsographical curve of basin

The coefficient of forestation K_1 may be obtained from the equation

$$K_1 = \frac{A_1}{A}, \qquad (4.38)$$

where A_1 = the forest area in the basin [km^2].
For the basin in *problem 10*

$$K_1 = \frac{0.75}{1.730} = 0.43 = 43 \text{ per cent.}$$

The average elevation of the basin above sea level is best computed from a *hypsometric mass curve*. This curve (*Fig. 4.9*) may be plotted in the following way: the basin elevations (according to the contour lines) are plotted as abscissae on graph paper, the area encompassed by the contour lines as ordinates of the graph. The points of the graph connected by a line represent the hypsometric curve of the basin. This is a mass curve of basin areas above sea level. If the area encompassed by the hypsometric curve and the co-ordinates is transformed into a rectangle whose base is A (total area of basin) the other dimension of this rectangle will be equal to the average elevation of the basin. For the basin in *problem 10*, a 20 m interval of contour lines was used and, from the hypsometric curve (*Fig. 4.9*), the average elevation was computed:

$$h_{az} = 386.6 \text{ m above sea-level (a.s.l.)}$$

50 per cent of the basin (median) lies below $h_{av} = 387$ m a.s.l.

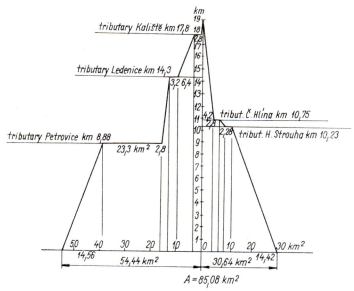

Fig. 4.10. Catchment area incremental graph

For a good representation of the *basin growth*, a graph may be plotted, starting at the spring and proceeding to the mouth. In *Fig. 4.10*, such a graph is plotted for a small basin in Czechoslovakia. Its construction is self-evident from the Figure.

4.5 Basic hydrometeorological characteristics of a basin

In order to compute the above characteristics all meteorological stations in the basin or in its vicinity are located. The meteorological service either publishes the data from these stations in a yearbook, or they may be requested from the service in their original form.

(a) Long-term precipitations

From these data is usually calculated:

Long-term and average annual precipitation at individual stations and space average for the basin.

Annual precipitation for individual years, months or seasons, their probability of occurrence p per cent (or return interval of N years) both for wet $(0 < 50\,p < \text{per cent})$ and for dry $(50 < p < 100\,\text{per cent})$ periods. These data may be computed for one meteorological station decisive for a given basin or region, or for space averages.

Daily, monthly, and annual evaporation from water surfaces in the basin. (For measurement of these data, *see page 58*.)

In addition, it is often necessary to compute the depth-duration-frequency-area relationships for storm rainfalls. These may be computed for the nearest meteorological station equipped with a recording rain-gauge, or by extrapolation techniques for stations with non-recording gauges and for areal averages (depth-duration-area curves).

For some hydrological problems, particularly for basin water balances, it is necessary to ascertain the values of the evapotranspiration from the basin for different periods of time. Evaporation from free water surfaces must be ascertained mainly for reservoir studies.

(b) Precipitation average for the basin

All meteorological stations in the basin and its immediate vicinity are used in determining the average precipitation over the basin.

For large basins with a sufficient density of precipitation gauges, *detailed isohyets* are drawn. Linear interpolation techniques are used in this drawing; nevertheless, an experienced hydrometeorologist's super-vision is most important in order to avoid errors of interpretation. The principle of increase of precipitation with geodetic elevation is often used to supplement the precipitation data; the isohyet pattern for long-term averages follows, in general, the contour-lines pattern of a basin. Exceptions are, however, frequent.

Isohyets may be drawn for long-term average, annual, and monthly precipitation, for individual years and months, and for isolated storms.

Using the isohyets, a space average over the basin may be computed by several methods. These have recently been described by Rainbird (1968).

The isohyet methods are generally used on large basins, however. They are too time-consuming for smaller basins and would not be accurate enough if only a few stations exist in the basin. Thus, for smaller basins, spatial average of precipitation is computed best directly from data of individual stations without the use of isohyetal maps. Before deriving this average, the data of all stations must be checked and often extrapolated. It is desirable that the period covered by the data be the same for all stations from which data are used.

By correlating to a series of data of a more reliable or longer observing station, data series of stations observing for a shorter period may be extrapolated, supplemented or simply checked, station by station, to see whether all the observations are complete and reliable. In a mountainous region, this correlation may only be used for two nearly adjacent stations. The coefficient of correlation, computed as indicated on *page 139*, may be a useful index of reliability of such extrapolation of data.

In less rugged terrain, on plains, *the double-mass analysis method* can be used for this purpose.

The principles of this method are as follows.

A certain number of stations (five at the minimum) with reliable observations of approximately the same length and in the same climatic region are selected as base stations.

The precipitation data (annual, monthly, seasonal) are added (accumulated) for each station, beginning with the last calendar observation. An arithmetic mean is calculated for the individual totals of the same calendar period. A similar cumulative addition of data from the station to be checked against the base stations is computed. Figures for corresponding time periods are plotted in rectangular co-ordinates in a double-mass curve (*Fig. 4.11*). If the data of the checked station are homogeneous with the data from base stations, the double-mass curve is a nearly straight line. If there is a break in the line at a certain point (*in Fig. 4.11* this is point *Z*), from this point into the future, observations from the station in question are either inaccurate or non-homogeneous. The non-homogeneity is to be investigated at the station in question since a climatic deviation would show up in the basic stations too. This method may also be used to check individual base stations, to eliminate those which are less reliable and thus attain, step by step, higher reliability of data.

This method allows for adjustment of data for unreliable periods. The correction to be applied is computed (*see Fig. 4.11*) as follows:

$$H_z = \frac{\tan \alpha}{\tan \alpha_0} H_0, \qquad (4.39)$$

where H_z is the adjusted precipitation (annual, monthly, seasonal),
$\quad H_0 =$ observed precipitation,
$\quad \tan \alpha =$ slope before change,
$\quad \tan \alpha_0 =$ slope after change.

In a similar way, data for individual stations may be supplemented or extrapolated.

In *problem 11*, the tabulated approach to the solution of the double-mass analysis is given. The entire computation is, however, omitted, in view of the considerable space required for it.

Problem 11

The observed annual precipitations for stations A, B, C, D, E, and F, are given for the following periods:

> A: 1930 – 1962,
> B: 1930 – 1962,
> C: 1928 – 1962,
> D: 1928 – 1962,
> E: 1930 – 1960,
> F: 1930 – 1962. (Observations missing
> for the period 1948 – 1955)

Check and extrapolate annual precipitations at station F, whose observation period covers 1930 – 1948 and 1955 – 1962, using the double-mass analysis method.

The calculations are set out as in *Table 18*.

The accumulated average of stations A, B, C, D, E, and F is plotted against the accumulated precipitation in F in *Fig. 4.11* using the last two columns of *Table 18*.

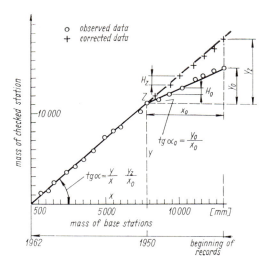

Fig. 4.11. Double mass curve

The data of station F may subsequently be extrapolated or adjusted, according to *Fig. 4.11*.

After all data from meteorological stations in the basin and its vicinity have been checked, possibly adjusted and extrapolated so that a homogeneous period of observation is available, *an average precipitation over the basin* may be computed. The simplest areal average is an arithmetic mean of station values. In many instances, this is sufficiently precise, particularly if the gauge network is not very dense. In most cases, sampling and instrument catch errors are very much more likely than computational errors.

Table 18

Year	A Obs.	A Cor.	B Obs.	B Cor.	C Obs.	C Cor.	D Obs.	D Cor.	E Obs.	E Cor.
1962	620		580		640		565			630
1961	560		530		600		515			570
1960	580		540		610		530		600	
⋮										
1930	650		600		670		590		620	

Year	F Obs.	F Cor.	∅ (A, B, C, D, E, F) Obs.	∅ (A, B, C, D, E, F) Cor.	Σ ∅ (A, B, C, D, E, F) Obs.	Σ ∅ (A, B, C, D, E, F) Cor.	ΣF Obs.	ΣF Cor.
1962	680		617	619	617	619	680	
1961	620		565	566	1 182	1 185	1 300	
1960	640		583	583	1 765	1 768	1 980	
⋮								
1930		710	635	640	37 866	37 243	38 235	38 739

When the network is non-homogeneous in space (unevenly distributed gauges), the '*Thiessen*' *polygon method* may be used.

This method provides for a weighting factor for each gauge. The stations are plotted on a map and connected by straight lines. The perpendicular bisectors of these connecting lines form polygons around each station. The sides of each polygon and the watershed divide represent the boundaries of the station area. The area of each polygon A_i is determined by planimetry (and may be expressed as a percentage of the total area). Weighted average precipitation P_0 for the basin is computed as the sum of the precipitations P_i at each station multiplied by its assigned weight (area A_i) and divided by the total area A of basin:

$$P_0 = \frac{\Sigma A_i P_i}{A}. \tag{4.40}$$

If A_i is expressed as a percentage of basin area eq. (4.40) will be

$$P_0 = \frac{\Sigma A_i P_i}{100}. \tag{4.41}$$

The following problem illustrates the procedure.

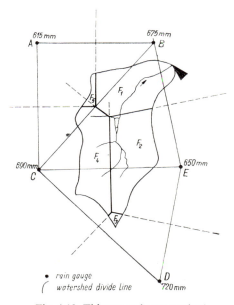

Fig. 4.12. Thiessen polygon method

Problem 12

Five stations, marked on the map in *Fig. 4.12*, may be used to compute the areal average precipitation in the basin from *problem 10*. The long-term annual average rainfall in these stations is: A = 615 mm; B = 675 mm; C = 690 mm; D = 720 mm; E = 650 mm. Compute this average, using the Thiessen polygon method.

The construction of polygons is self-evident from *Fig. 4.12*. To determine the weight of the stations, the average planimeter reading $f(A_i)$ is noted:

$$f(A_1) = 0.150, \quad \text{or} \quad 32.6 \text{ per cent } A,$$
$$f(A_2) = 0.166, \quad \text{or} \quad 36.1 \text{ per cent } A,$$
$$f(A_3) = 0.005, \quad \text{or} \quad 1.1 \text{ per cent } A,$$
$$f(A_4) = 0.133, \quad \text{or} \quad 28.5 \text{ per cent } A,$$
$$f(A_5) = 0.008, \quad \text{or} \quad 1.7 \text{ per cent } A,$$

$$0.462 \qquad\qquad 100.0 \text{ per cent } A.$$

Substituted in equation (4.41), this gives:

$$S_0 = \frac{615 \times 32.6 + 675 \times 36.1 + 720 \times 0.1 + 690 \times 28.5 + 650 \times 1.7}{100} =$$

$$= 660 \text{ mm}.$$

The arithmetic mean of precipitation in the basin is

$$S_0 = \frac{615 + 675 + 720 + 690 + 650}{5} = 668 \text{ mm}.$$

The difference between the Thiessen polygon and the arithmetic mean is relatively small in this example.

(c) Annual, monthly, and seasonal precipitation of a given return interval

Annual, monthly, or seasonal precipitation from individual stations for which a certain return interval has been computed are needed, mainly, for irrigation and drainage projects. A frequency distribution curve (or a probability curve), plotted directly from the data, may be used for return intervals equal or slightly larger than the observation period (in years). Theoretical frequency distribution curves (in Czechoslovakia, the Type III Pearson curve) are used to extrapolate to longer return intervals.

The procedure is identical to the one described for annual precipitation in *problems 7 and 8 (pages 155 and 162)*. It is to be noted that frequencies of averages of individual station values may not correspond to the frequencies of areal means of the same magnitude of the event.

(d) Intensity of storm rainfall

Another important hydrometeorological element is the intensity of storm rainfall of various durations and various frequencies.

If there are recording raingauge data on rain intensities for at least ten years, an intensity series may be plotted from an analysis of the records. Different methods may be used for the plotting of such series; in particular a distinction is made between partial duration and annual series (*see pages 149 and 159*). It is recommended that partial duration series be used for this purpose, since the return interval required for the use of most rain intensity data is less than 5 years.

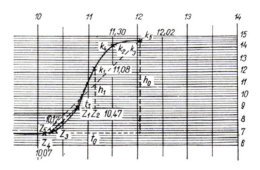

Fig. 4.13. Rainfall recording chart

The following method (that of Reinhold) is used in central and eastern Europe (Germany, Austria, Hungary, and, in particular, in Czechoslovakia).

For design purposes, data are needed on rainfalls of various durations and of various frequencies whose intensity is constant during the whole duration. Real rainfall is of varying intensity, however, usually being low at the beginning and end, and higher, sometimes with several peaks, in between. This rainfall is to be replaced by one or more rainfall intervals with a constant intensity and various durations, which are equivalent, having the same effect as natural rainfall. This is obtained by an analysis of the rain-recorder chart. The Reinhold method, which avoids the subjective error, departs from the observation that storm rainfalls may be simple, double, and multiple. Simple rainfall is that in which the intensity reaches only one peak. A record of this type of rainfall is shown in *Fig. 4.13*. It is a rainfall mass curve in time; the time is on the ordinates and the rain depth on the abscissae. The intensity at each point of the record is the value of the tangent at this point to the curve; the average intensity of the entire rainfall is the value of the tangent (slope) of the straight line joining the beginning and end of the rainfall. In *Fig. 4.13*:

$$i_{av} = \frac{h_0}{t_0}.$$

A simple rainfall is analysed by the following procedure. Minor inequities in the registered curve are adjusted, the points in which the curve breaks are identified, and the curve is replaced by several straight lines joining these points. In this way, the entire rainfall is divided into rain intervals with constant intensities. A transparent overlay can help in finding the break points, a narrow straight-line slit in it being moved along the recorded curve.

The highest intensity (the steepest part of the record) is identified and computed by measuring t and h. In *Fig. 4.13* it will be $i_1 = \frac{h_1}{t_1}$. The interval with the next highest intensity is then identified. Its duration and depth of rain in millimetres are measured and added to the first, and the intensity is calculated as follows:

$$i_2 = \frac{h_1 + h_2}{t_1 + t_2}.$$

This procedure is repeated until the interval of lowest intensity is reached. In this way, a series of decreasing intensities and of increasing durations is obtained.

Problem 13a

Analyse the storm recorded in *Fig. 4.13* and derive from it data for a partial duration series of intensities.

The rainfall started at 10.07 and ended at 12.02; its total duration was 115 min and its total depth was $14.6 - 6.9 = 7.7$ mm. The breaks of record were connected by straight lines. The interval with the highest intensity is marked z_1k_1, its duration is $t_1 = 11.08 - 10.47 = 21$ min, its depth is 3.4 mm, and thus its intensity is $i_1 = i_{21\,min} = \dfrac{3.4}{21} =$ $= 0.162$ mm/min (or 27 l/s ha). The second steepest interval is marked k_1k_2; it lasted 22 min and 1.8 mm of rain fell during this time. Both periods of rain duration and depth are added:

$$i_{43\,min} = \frac{3.4 + 1.8}{21 + 22} = \frac{5.2}{43} = 0.121 \text{ mm/min (20.21 l/s ha).}$$

Similarly,

$$i_{73\,min} = \frac{7.1}{73} = 0.0973 \text{ mm/min (16.2 l/s ha)}$$

and for the entire rainfall

$$i_{115\,min} = \frac{7.7}{115} = 0.0669 \text{ mm/min (11.15 l/s ha).}$$

(The conversion of mm/min to l/s ha is most usual in central Europe; 1 mm/min $= 166.67$ l/s ha.)

Thus, the whole simple rainfall has been analysed and a series of five equivalent rain intervals of constant intensity has been obtained.

A multiple rainfall is one whose intensity has several peaks; two simple rainfalls following each other, for instance, are a double rainfall and they are evaluated consecutively in a similar way to that shown above. Triple rainfall is divided into simple rainfalls I, II, and III and these are then analysed in combinations (I and II), (III and I), (II and III), each being evalu-

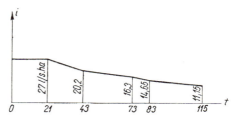

Fig. 4.14. Equivalent rainfall curve

ated as simple rainfalls. From these groups, the one which has the highest intensities for the same duration is retained. This procedure is repeated until the entire multiple rainfall is analysed like a simple rainfall.

The intensities for different durations are plotted in a graph (*Fig. 4.14*). The graph can also be plotted in logarithmic scales. A curve of equivalent rain intensity-duration is thus obtained.

From this curve may be abstracted equivalent intensities for selected durations. The durations usually used in central Europe are: 5, 10, 15, 20, 30, 60, 90, 100, and 120 min. In this manner, rainfalls which have occured over several years may be evaluated (for reliable extrapolation, more than 10 years of observations are necessary), a partial duration series of intensities may be obtained for each duration of rainfall and a frequency analysis of the series computed. Annual maximum series may also be computed for this purpose.

In several countries of the world, in particular in the USA, frequency analyses of depths of rainfall for longer durations, such as 6, 12, 24, 48, and 72 h, are also computed.

For frequency analysis of intensities of durations less than 2 h, and return intervals of less than 2 years, the use of partial duration series is, however, strongly recommended. For frequency analysis of depths of rainfall of more than 2 h, and for return intervals above 5 years, annual maximum series may be used with confidence. The conversion table (*see page 159*) of annual to partial duration series may also be used.

Several approaches may be adopted for the frequency analysis. They have been basically exposed in *chapter 3* and in *problems 6, 7, 8*, and *9*.

For this reason, only procedures specifically used for rainfall series will be briefly exposed here.

For short return intervals up to 5 years, the use of a suitable probability paper is sufficient. The plotting positions of variates are computed by the formulae

$$T = \frac{n+1}{m} \quad \text{or} \quad T = \frac{n+0.4}{m-0.3},$$

where T = return interval,

n = number of years of record,

m = rank of the variate in annual series, arranged in the descending order.

For longer return period and extreme values of rain intensity or rain depth, the use of a theoretical frequency distribution is needed for extrapolation of the series.

One of the most used of such distributions is the extreme value (Gumbel) distribution. For its computation, it is particularly advantageous to use the method of the frequency factor K (Chow, 1951), using the equation

$$x = \bar{x} + \sigma K,$$

where x = extrapolated variate,
    $\bar{x}$ = arithmetic average of the series,
    σ = standard deviation of the series,
    K = frequency factor, which is a function of the recurrence interval and of the type of frequency distribution, found in tables.

It is to be noted that, if the above equation is divided by $\bar{x}$, the equation used on *page 165* in connection with the Pearson Type III distribution curve is obtained.

$$\frac{x}{\bar{x}} = \frac{\bar{x}}{\bar{x}} + \frac{\sigma K}{\bar{x}} = k_{p\%} = 1 + C_v K. \qquad (4.27)$$

where $k_{p\%}$ = extrapolated reduced variate (to the mean),
    C_v = coefficient of variation.

To compute the frequency factor for the Gumbel extreme value distribution, eq. (4.27) may be transformed with respect to the value of the reduced variate of the mode:

$$K = \frac{Y_T - \bar{Y}_n}{S_n},$$

where Y_T = reduced variate as a function of return interval T. For Gumbel distribution,

$$Y_m = -\left(0.834 + 2.303 \log \log \frac{T}{T-1}\right),$$

    S_n = reduced standard deviation, a function of n (size of sample),
    $\bar{Y}_n$ = reduced mean (to mode) depending only on the size of the sample n.

The values of S_n, $\bar{Y}_n$ and Y_T (for return intervals of 2, 5, 10, 25, 50, and 100 years) are included in *Tables 19a, b,* and *c*.
The computation is illustrated in *problem 13b*.

Problem 13b

Included in *Table 19d* are annual maxima of rainfall intensities of 30 minutes duration, recorded and analysed for the period from 1943 to 1956. Compute, using Gumbel extreme values distribution (double exponential), and the frequency factor procedure, intensities of recurrence intervals of 2, 5, 10, 25, 50, and 100 years.

The preliminary computation of the arithmetic mean and the standard deviation are in lines 2 and 3 of *Table 19d*.

Table 19a. Reduced mean Y_n

m	0	1	2	3	4	5	6	7	8	9
10	0.4952	0.4996	0.5035	0.5070	0.5100	0.5128	0.5157	0.5181	0.5202	0.5220
20	0.5236	0.5252	0.5268	0.5283	0.5296	0.5309	0.5320	0.5332	0.5343	0.5353
30	0.5362	0.5371	0.5380	0.5388	0.5396	0.5402	0.5410	0.5418	0.5424	0.5430
40	0.5436	0.5442	0.5448	0.5453	0.5458	0.5463	0.5468	0.5473	0.5477	0.5481
50	0.5485	0.5489	0.5493	0.5497	0.5501	0.5504	0.5508	0.5511	0.5515	0.5518
60	0.5521	0.5524	0.5527	0.5530	0.5533	0.5535	0.5538	0.5540	0.5543	0.5545
70	0.5548	0.5550	0.5552	0.5555	0.5557	0.5559	0.5561	0.5563	0.5565	0.5567
80	0.5569	0.5570	0.5572	0.5574	0.5576	0.5578	0.5580	0.5581	0.5583	0.5585
90	0.5586	0.5587	0.5589	0.5591	0.5592	0.5593	0.5595	0.5596	0.5598	0.5599
100	0.5600									

Table 19b. Reduced standard deviation S_n

m	0	1	2	3	4	5	6	7	8	9
10	0.9496	0.9676	0.9833	0.9971	1.0095	1.0206	1.0316	1.0411	1.0493	1.0565
20	1.0628	1.0696	1.0754	1.0811	1.0864	1.0915	1.0961	1.1004	1.1047	1.1086
30	1.1124	1.1159	1.1193	1.1226	1.1255	1.1285	1.1313	1.1339	1.1363	1.1388
40	1.1413	1.1436	1.1458	1.1480	1.1499	1.1519	1.1538	1.1557	1.1574	1.1590
50	1.1607	1.1623	1.1638	1.1658	1.1667	1.1681	1.1696	1.1708	1.1721	1.1734
60	1.1747	1.1759	1.1770	1.1782	1.1793	1.1803	1.1814	1.1824	1.1834	1.1844
70	1.1854	1.1863	1.1873	1.1881	1.1890	1.1898	1.1906	1.1915	1.1923	1.1930
80	1.1938	1.1945	1.1953	1.1959	1.1967	1.1973	1.1980	1.1987	1.1994	1.2001
90	1.2007	1.2013	1.2020	1.2026	1.2032	1.2038	1.2044	1.2049	1.2055	1.2060
100	1.2065									

Table 19c. Return period as a function
of reduced variate

Return period [years]	Reduced variate
2	0.3665
5	1.4999
10	2.2502
25	3.1985
50	3.9019
100	4.6001

Table 19d

Year	Intensity [mm/min]	m	x [mm/min]	x^2
43	0.212	1	1.630	2.6569
44	0.483	2	1.240	1.5376
45	0.671	3	0.863	0.7448
46	0.363	4	0.850	0.7225
47	0.526	5	0.671	0.4502
48	0.580	6	0.580	0.3364
49	0.863	7	0.526	0.2767
50	0.316	8	0.483	0.2333
51	1.630	9	0.463	0.2144
52	1.240	10	0.364	0.1325
53	0.850	11	0.316	0.0999
54	0.151	12	0.300	0.0900
55	0.300	13	0.212	0.0449
56	0.463	14	0.151	0.0228
			Σ 8.649	Σ 7.5609

The standard deviation is computed by a short-cut equation

$$\sigma = \sqrt{\frac{\Sigma(x)^2 - \bar{x}\,\Sigma x}{n - 1}}$$

Table 19e. Rainfall intensity-frequency analysis

Station __M__   Computed by __Novak__   Date __20. 9. 69__

Period of record __1943—1956__   Checked by __Novotny__   Date __25. 9. 69__

Line Operation	5 min	10 min	15 min	30 min	1 h	2 h	24 h
1. Σx				8.649			
2. n = number of items				14			
3. $\bar{x} = (1)/(2)$				0.6177			
4. Σx^2				7.5609			
5. $\bar{x}\,\Sigma x = (1) \times (3)$				5.3425			
6. $\Sigma(x - \bar{x})^2 = (4) - (5)$				2.2174			
7. $\sigma^2 = (6)/(n - 1)$				0.1710			
8. $\sigma = \sqrt{(7)}$				0.414			
9. S_n (*Table 19b*)				1.0095			
10. Y_n (*Table 19a*)				0.5100			
11. $1/a = (8)/(9)$				0.413			
12. $Y_n/a = (10) . (11)$				0.2106			
13. $n = (3) - (12)$				0.4071			
14. $(11) \times 0.3665$ (*Table 19c*)				0.1516			
15. $(11) \times 1.4999$ (*Table 19c*)				0.6200			
16. $(11) \times 2.2502$ (*Table 19c*)				0.9293			
17. $(11) \times 3.1985$ (*Table 19c*)				1.3216			
18. $(11) \times 3.9019$ (*Table 19c*)				1.6087			
19. $(11) \times 4.6001$ (*Table 19c*)				1.8998			
20. $x_2 = (13) + (14)$				0.5587			
21. $x_5 = (13) + (15)$				1.0271			
22. $x_{10} = (13) + (16)$ [mm/min]				1.3264			
23. $x_{25} = (13) + (17)$				1.7287			
24. $x_{50} = (13) + (18)$				2.0158			
25. $x_{100} = (13) + (19)$				2.3062			

and x_T or $(x_2, x_5, x_{10}, x_{25}, x_{50},$ and $x_{100})$ are computed from the equation

$$x_T = \bar{x} + \frac{Y_T - \overline{Y_n}}{S_n}\,\sigma = \bar{x} - \frac{\overline{Y_n}\sigma}{S_n} + \frac{Y_T\sigma}{S_n}.$$

If we substitute $\dfrac{S_n}{\sigma} = a$ or $\dfrac{\sigma}{S_n} = \dfrac{1}{a}$ and $n = \bar{x} - \dfrac{\overline{Y_n}\sigma}{S_n} = \bar{x} - \dfrac{\overline{Y_n}}{a}$ the

final equation used is $x_T = n + \dfrac{Y_T}{a}$, Y_n, S_n, and Y_T are taken from

Tables 19a, b, and *c* respectively assuming $m = 14$ and $T = 2, 5, 10, 25, 50,$ and 100.

The computation is made in a routine table where different operations are indicated step by step on each line (*see Table 19e*). Lines $20-25$ are the solution of *problem 13b*. The computed points are plotted on the extreme probability paper (Gumbel type); they form a straight line (*Fig. 4.15*).

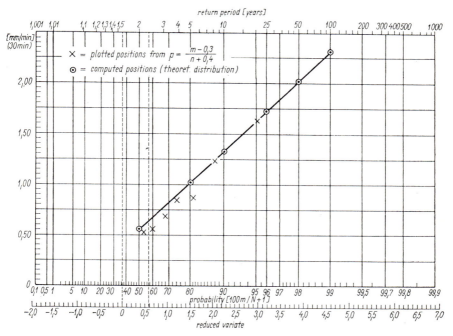

Fig. 4.15. Gumbel extreme distribution on extreme probability paper

The original data of *Table 19d*, plotted on probability paper, using plotting positions computed by the equation

$$T = \frac{n + 0.4}{m - 0.3},$$

indicate that the extreme values extrapolated by the Gumbel theoretical distribution ($T = 25, 50, 100$ years) fit well with the original series.

In view of the straight line indicated in *Fig. 4.15*, computations may be considerably shortened if values for durations 1 and 24 hours are computed only for $T = 2$ and 100 years, and the rest of the values are interpolated graphically by plotting on a probability paper.

Confidence interval

Extrapolated values of x_T may involve a considerable sampling error. The confidence interval for these values is

$$x_T \pm t(a) \, S_e,$$

where $a = $ confidence probability,
$\quad$ t(a) = function of a,
$\quad\quad S_e = $ probable error (deviation).

For

$$a = 90 \text{ per cent } t(a) = 1.645,$$
$$a = 80 \text{ per cent } t(a) = 1.282,$$
$$a = 68 \text{ per cent } t(a) = 1.000,$$

and

$$S_e = b \frac{\sigma}{\sqrt{m}},$$

where $b = \sqrt{1 + 1.3K + 1.1K^2}$ (for Gumbel distribution),
and $\quad K = $ frequency factor for a given T,
$\quad\quad m = $ number of years of record (or annual variates)

For the value of x_{100} from *problem 13b*, the confidence interval of 90 per cent probability is $x_{100} \pm 1.645 \, S_e$

$$S_e = 4.762 \frac{0.414}{3.7417} = 0.3075,$$

thus x_{100} has the confidence interval 2.3062 ± 0.5060.

Absence of data

In the absence of enough data, extrapolations may be obtained on the basis of the relation between the incidence of thunderstorms or thunderstorm days and the ratio of 24-hour rainfall to 1-hour rainfall. In Czechoslovakia, such a relationship has been established by Trupl for different intensities, while Hershfield (1957) has ascertained the following relationship from studies based on a wide range of climates:

Ratio of 1-hour to 24-hour rainfall	0.2	0.3	0.4	0.5
Mean annual incidence of thunderstorms (days)	1	8	16	24

Curves and formulae for rainfall intensity, duration and frequency

The data ascertained by the above procedures may be plotted for ready use, either into graphs (curves of rainfall intensity, duration, and frequency), or expressed in formulae for which local parameters are tabulated. In Europe, the graphs generally have the form shown in *Fig. 4.15a*.

The formulae can be expressed in the general form

$$i = \frac{KN^a}{(t + b)^n} \quad \text{or} \quad i = \frac{K \log N}{(t + b)^n},$$

where i = equivalent intensity of rainfall of a duration t and a recurrence interval of N years,

K, a, b, and n = parameters of geographical (climatic) influence.

Using extrapolation by the Goodrich distribution curve, the author of this book derived the following formula (Němec, 1957)

$$hP = (a \log t + b) N^n, \tag{4.42}$$

where hP = depth of rainfall [mm] of duration t [min] and recurrence interval N [years],

a, b, n = parameters of local influence (these parameters were ascertained for the entire territory of Bohemia by the author).

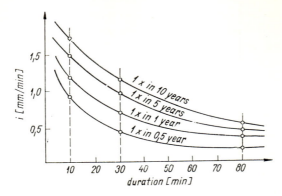

Fig. 4.15a. Curves of rainfall intensity, duration and frequency

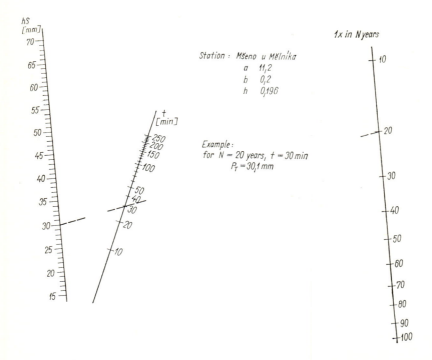

Fig. 4.15b. Nomogram for equivalent intensity of rainfall

The formula may be transformed into a nomogram (*Fig. 4.15b*). It is valid for durations of 5 to 120 min, and also for 6-, 12-, and 24-h rainfalls. The values given by the formula for these longer durations of rainfall are slightly higher than the actual observed data but, in the absence of such observed data, the computed values include a safety margin.

Depth-area studies

Rainfall-frequency analyses are generally made for point or station data. Care must be exercised to ascertain that the station data used are representative of the area for which the results of the frequency analysis are to be used. In flat regions, data from almost any station should be representative, provided the record is long enough and the gauge is properly exposed. In mountainous regions, however, the effective area of representation is very small.

In some countries, depth-area curves have been constructed for reducing point rainfall-frequency data for areas up to 1000 km². However, the density of raingauge networks is generally inadequate to derive a universally valid set of these curves. In Czechoslovakia, a very approximative consideration of Reinhold is used for small basins up to 25 km². According to Reinhold, the point rainfall intensity is to be reduced on a basin of area 10 km² by 5 per cent and for 25 km² by 10 per cent.

Depth-area-duration studies

For large basins in which rainfall depths during 6, 12, 24, 48, and 72 h are decisive, the technique devised by the US Weather Bureau may be used advantageously.

A most detailed description and step-by-step procedure for the depth-area-duration (DAD) studies is included in the WMO Manual on Depth-Area-Duration Analysis of Storm Precipitation (WMO publication No. 237 TP 129, 1969, WMO Geneva). Manual as well as computer techniques are included in this publication; therefore, only a short outline of this technique is included below.

The first step in a DAD study is to plot accumulated values of rainfall against the time of day (a mass curve or integral curve) for each station, or for selected stations if there are a great many. The mass curves for non-recording stations are constructed by comparison with mass curves from recording stations, by means of proportionality factors, taking into account the movement of the storm, reports of times of beginning and ending, and reports on times of heaviest rainfall.

The pertinent stations are then listed in a table, and accumulated values of rainfall are tabulated for each station, using 6-h increments. Other time increments could be used, but 6 h has been found to be most practical for the majority of purposes. For convenience, the stations are listed in order of decreasing magnitude of total storm rainfall.

The next step is to examine the table and select the particular 6-h period which has the largest of the rainfall increments. The values for this time increment are then listed. The period of maximum 12-h rainfall is found similarly and its rainfall listed. The same operation is applied to obtain the maximum 18-h, 24-h, and further 6-h increments. For periods embracing several 6-h increments, considerable trial and error may be necessary to find the period which includes the maximum rainfall for the particular duration.

From the tabulation of maximum $n-h$ rainfall increments ($n = 6$, 12, 18, 24, 48, 72), isohyetal maps are prepared for each duration: 6 h, 12 h, etc. These maps are then evaluated by use of a planimeter or by tallying grid points, and the resulting values are plotted on a graph of area against depth, with a curve for each duration. Commonly, the depth scale is a linear ordinate, and area is the abscissa with a log scale.

Probable maximum precipitation (PMP)

For flood design studies, particularly to ascertain the value of rainfall considered as a physical maximum that may occur over a basin, the so-called 'probable maximum precipitation' (PMP) may be computed. The procedures are based on meteorological considerations and models of moist air motion are devised for this purpose. Two such models are considered: the convergence model and the orographic model. A third approach ascertains the PMP on the basis of a statistical-empirical method from observed maximum values of 24-h point rainfall. For the UK, this last method was used by Rodda (1969).

The meteorological approaches (models of air motion) are advantageously combined with observed storm transposition and storm maximization.

All these techniques require considerable meteorological knowledge and should be applied only by experienced hydrometeorologists. Their detailed description is included in the WMO Technical Note No. 98 'Estimation of Maximum Floods' (WMO, No. 233. Tech. Publication 126, WMO Geneva, 1969) and the reader is directed for references and further guidance to this publication.

(e)   Evaluation of snowpack and snowmelt computation

In addition to liquid precipitation, it is often necessary to evaluate solid precipitation, i.e., snow. This generally involves calculating the *water equivalent of the snow* and the *snowmelt*. The water equivalent is most important at the end of the winter, that is, at the beginning of thaw. On the basis of measurements of snow depth and its water equivalent at stations or along snow courses (*see page 83*), the water storage in the snow [mm] is determined by the following equation

$$S_z = dS_s, \tag{4.43}$$

where S_z = depth of water storage in the snow (retained liquid precipitation),
S_s = depth of snow cover,
d = water equivalent (density) of snow.

If the snow density (water equivalent) has not been measured, it can be roughly estimated at 0.1 for snow at the beginning of the winter and at 0.3 at the end of the winter or according to *Table 20*.

In using *Table 20*, it should be kept in mind that the snow cover is generally made up of several layers.

From the water equivalent thus obtained at individual stations or in snow courses, the average water equivalent of the snow pack throughout the entire basin can be derived by the same method which was used for deriving average rainfall in a basin (the Thiessen polygon method, for instance). An average is most accurate, however, if the snow courses

Table 20

Type of snow	Water equivalent of snow (density)	Type of snow (in mountains)	Water equivalent of snow (density)
Fresh powder snow	0.05	Packed snow	0.20
		Granulated snow	0.25
Ordinary fresh snow	0.10	Coarse snow (firn)	0.5
Virgin snow	0.15	Granular ice	0.85
		Glacier	0.90

cover the entire basin or if the snow is measured in a rectangular grid over the basin.

Nevertheless, in most cases it is necessary to know not only the total volume of water stored in the snowpack (the total snowmelt volume) but also the timing of the melt and runoff from the snowpack. This is a very complicated problem and it can be treated basically in two ways. The simpler one originated and is being widely used with success in the United States; a similar method also exists in the Soviet Union (G. A. Alexejev). It is the *degree-day method*, and has the advantage of using generally accessible meteorological data. The basis of this method is the equation

$$S = kD, \tag{4.44}$$

where S = the depth of water which melts in one day,

$\quad\quad k$ = coefficient expressing the influence of natural and climatic conditions in the basin on snowmelt,

$\quad\quad D$ = positive daily air temperature, for the day for which S is being determined, calculated in the average height of the basin.

The average or maximum daily temperature may be used. The value of k being empirical and often transposed from the USA where it is computed in °F and inches a conversion factor for °C and cm may be used. The equation will then be:

$$S \ [\text{cm}] = 4.572kD \ [°\text{C}], \tag{4.45}$$

where k is derived for °F and in.

Coefficient k integrates all influences (except for temperature) which may have an effect on snow melt and the runoff of snow water. These influences differ not only from one basin to another but also from day to day, and therefore only average values may be derived. In the United States, the value of $k = 0.06$ is often used.

The values of k with respect to basin conditions are given in *Table 21* (in the second column k is given in cm/°C).

Table 21

Conditions of snow melt	k [original units: in/°F]	k [cm/°C]
Very low melt	0.02	0.09
Heavily forested region or bared slopes with northern exposure	0.04—0.06	0.18—027
Average value	0.06	0.27
Forested slopes with southern exposure or unforested regions with average melt	0.06—0.08	0.27—0.36
Very heavy melt	0.30	1.8

The Hydraulic Research Institute in Prague measured the coefficient k with the aid of a cobalt 60 radio-isotope. It was found that k varies with snow density and wind speed. The relation represented in *Fig. 4.16* was obtained at one station in the Krkonoše Mountains of Czechoslovakia. This graph also shows that the value 0.27 (0.06 in the original units) is indeed an average since the snow density of 20—30 % is most common.

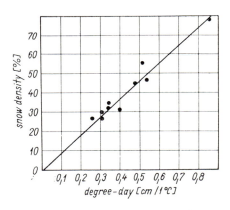

Fig. 4.16.

Average daily temperatures for the mean elevation of the basin should be used. If the basin is steep (as is usual in mountain regions), it should be kept in mind that with increased elevation the air is cooled, roughly by 0.7 °C per 100 m. If temperature observations come from stations whose elevations are different from the average elevation of the basin, it is advisable to convert them to this elevation using equation

$$T_0 = T_p - \frac{0.7}{100}(h_{mean} - h_p), \qquad (4.46)$$

where T_0 = the temperature adjusted to the average elevation of the basin (h_{mean}),

T_p = temperature observed at station with elevation h_p.

If there are more stations in the basin, the arithmetical average is calculated from the adjusted temperature.

Improvement of the accuracy of this method should consist of the following steps:

(a) a more exact determination of average snow depth in the basin (by detailed snow courses);

(b) more exact determination of average water equivalent of snow;

(c) only if greater precision has been achieved by steps (a) and (b), a more precise value of the coefficient k should be derived.

The 'degree-day' method is illustrated in the problem below. It is assumed that the sequence of daily temperatures in the melt period is known.

Problem 14a

The average daily air temperatures are given for stations A (elevation 820 m) and B (elevation 610 m). Using the degree day method, compute the melt of the snow pack of a depth of 125 cm and of an average water equivalent of 0.2 on a basin with an average elevation of 750 m. Adjust the observed temperatures at stations A and B to the average elevation of the basin.

For a temperature of 1 °C at station A

$$T_0^1 = 1 - \frac{0.7}{100}(750 - 820) = 1 + 0.49 \cong 1.5 \text{ °C,}$$

For a temperature of 2 °C at station B

$$T_0^2 = 2 - \frac{0.7}{100}(750 - 610) = 2 - 0.98 \cong 1 \,°C.$$

The average from stations A and B

$$T_0 = \frac{T_0^1 + T_0^2}{2} = \frac{1 + 1.5}{2} = 1.25 \,°C, \qquad \text{and so on.}$$

The computation of melt is indicated in *Table 22*. The procedure is as follows:

Table 22

Date	Average temperature in basin T_0 [°C]	Degree-days	Daily melt S [mm]	Depth of water left in snow [mm]
4. IV.	0	0	0	1250 × 0.2 = 250
5. IV.	1.7	1.7	4.7	245.3
6. IV.	1.1	1.1	3.1	242.2
7. IV.	2.2	2.2	6.2	236.0
8. IV.	8.9	8.9	24.9	211.1
⋮	⋮	⋮	⋮	⋮

At the beginning of melt, the total water storage in the snow is 1250 × × 0.2 = 250 mm (*Table 22*). Melt is calculated using eq. (4.45), the value of k being taken as 0.06. Thus, for 5 April,

$$S = 4.572 \times 0.06 \times 1.7 = 0.47 \,\text{cm} = 4.7 \,\text{mm}.$$

The computation proceeds in the same way until all the snowpack is melted.

The second and more sophisticated method for snow-melt computation is the use of the snow energy budget equation. A much larger amount of meteorological data is, however, necessary for the use of this method. In particular, net solar radiation, average basin forest canopy, albedo of the snow, wind speeds, average daily dewpoints, and psychometric measurements must be available.

This method (as well as the 'degree-day' procedure) are described in the WMO Guide to Hydrometeorological Practices.

Estimation of daily temperature sequence

For both the degree-day and the energy budget methods, the most critical sequences of daily air temperature must be estimated. For this purpose, curves of the highest maximum (or average) daily temperatures are developed for the snow-melt season on the basis of moderately long records (30 – 50 years) of air temperature from a representative station within the basin. Greatest recorded temperatures for durations of 3, 7, 15, and 20 days are abstracted from such records. Experience has shown that plotting of the highest 3- or 7-day temperatures gives curves in which the maximum recorded values are approached asymptotically. ('Estimation of Maximum Floods' WMO Technical note No. 98, page 131.) The curves are plotted in a graph of which the ordinate is the calendar time of the snow-melt season and the abscissae are temperatures of 3-, 7-, 15-, and 20-day series.

Once such curves are plotted, the sequence is selected so as to fall within the snow-melt period and yield maximum snow-melt. This usually means allowing for low temperatures during the end of the snowfall period, prohibiting premature snow-melt. Then the temperatures are allowed to increase rapidly to maximum values and remain that way for as long as the limiting curves permit.

For flood runoff studies, the possibly simultaneous occurence of snow-melt and heavy rain must always be considered. While the rainfall sharply increases the total volume of effective runoff, it reduces the snow-melt to a certain degree by lowering the 'degree-day' factor k or the net radiation, dew point temperature, wind, and insolation to be used in the energy budget method. The above-mentioned publications include detailed discussion of methods to be used for the simultaneous snow-melt and rainfall consideration in flood runoff computation.

(f)   Computation of evaporation

One of the most difficult tasks in hydrology is to estimate evaporation, either from free water surfaces or from soil and plants (evapotranspiration).

Evaporation from water surfaces

As has been pointed out on *page 61*, the measuring of evaporation from water surfaces has several shortcomings. In particular, areal estimates of evaporation (over a basin) based on point measurements of pan evaporimeters are unsatisfactory.

The point measurements of evaporation pans are somehow biased by microclimatical conditions and their extrapolation to large areas is therefore not always reliable. The zonal character of meteorological elements offers another approach to the problem of areal estimation of evaporation. In particular, the energy budget method combined with the aerodynamic effect, serves as a basis for several equations, of which Penman's formula is one of the best-known:

$$hE(\mathrm{d}) = \frac{\Delta H + E_a\gamma}{\Delta + \gamma},$$

where hE (d) = the daily evaporation [mm],

$\quad\quad\quad\Delta$ = the slope of the saturation vapour pressure curve at temperature T_a,

$\quad\quad\quad H$ = net radiation,

$\quad\quad\quad\gamma$ = psychrometric constant,

$\quad\quad\quad E_a$ = parameter including wind velocity and the saturation deficit.

Although the use of data from direct measurements for solar and long-wave radiation provides the best results for the application of this formula, the computation may be based on the following data:

$\quad t$ = mean daily temperature [°C],

$\quad \dfrac{n}{D}$ = ratio of actual to possible daily hours of sunshine,

$\quad R_A$ = mean extra-terrestrial radiation expressed in equivalent of mm/day evaporation,

$\quad h$ = relative humidity of air (daily average),

$\quad u_2$ = average daily wind velocity at 2 m [m/s].

The nomogram in *Fig. 4.17* permits a rapid computation of the daily values.

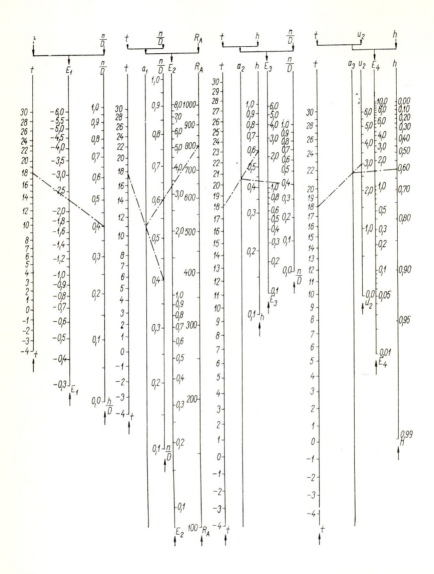

Fig. 4.17. Nomogram for computation of evaporation E_0 from free water surface according to Penman

Problem 14b

Using nomogram on *Fig. 4.17*, compute daily evaporation from free surface for the following data:

average daily temperature $\qquad t = 18\ ^\circ\text{C}$,

cloudiness ratio $\qquad \dfrac{n}{D} = 0.4$,

mean extraterrestrial radiation $\qquad R_A = 800$ g/cal cm^2 day,

mean daily relative humidity $\qquad h = 0.6$,

average daily wind speed at 2 m $\qquad u_2 = 3.0$ m/s.

The nomogram decomposes the computation into four values, E_1, E_2, E_3, E_4, to be added subsequently. According to the nomogram, $E_1 = -2.28$, $E_2 = +3.30$, $E_3 = +1.12$, $E_4 = 1.52$

$$E_1 + E_2 + E_3 + E_4 = 3.66 \text{ mm/day.}$$

Hence, daily evaporation for the above data is 3.66 mm/day.

It is to be noted that the radiation data R_A and the cloudiness ratio $\dfrac{n}{D}$ may be taken as an average from geophysical tables. Thus, the computation with Penman's equation may not require more directly measured data than other more empirical approaches. However, if there are measured data, Penman's approach yields more accurate results.

Simpler equations based on the Dalton law and on the correlation of evaporation with wind velocity are used extensively in several parts of the world. The Meyer formula, as adapted by Tichomirov, was used in the USSR in the form

$$hE \text{ (m)} = d(15 + 3w), \tag{4.47}$$

where hE (m) = the total monthly evaporation from free water surface [mm],

$\qquad d$ = average monthly saturation deficit [mm Hg],

$\qquad w$ = average monthly wind speed at weathervane height (10 m) [m/s].

A similar equation has been derived by the Hydraulics Research Institute in Prague for conditions in Bohemia and Moravia (J. Váša: Evaporation from Water Surfaces, Prague-Podbaba 1959):

$$hE \text{ (d)} = (e - e_a) (0.0405 \text{ B} + 0.364\ 15), \qquad (4.48)$$

where hE (d) = the average daily evaporation [mm],

e_s = tension of saturated vapours at surface temperature of the evaporating water [mm Hg],

e_a = tension of water vapours in air (at a height of 2 m),

B = wind speed in Beaufort degrees.

To facilitate the calculations, the values e_s and e_a may be found in the psychrometric table in *appendix 6* and the difference multiplied (by slide rule) by the value f (B) from *Table 23*. For calculating e_s, the temperature of the water surface must be known. For this purpose, a table of differences between average monthly temperatures of the water surface t_s and air temperature t_a during the months of the vegetation season was computed (*Table 24*).

Table 23

B	0.0	0.1	0.2	0.3	0.4	0.5	0.6	0.7	0.8	0.9	1.0
$f(B)$	0.36	0.37			0.38			0.39		0.40	
B	1.1	1.2	1.3	1.4	1.5	1.6	1.7	1.8	1.9	2.0	2.1
$f(B)$	0.41			0.42			0.43		0.44		0.45
B	2.2	2.3	2.4	2.5	2.6	2.7	2.8	2.9	3.0	3.1	3.2
$f(B)$	0.45	0.46			0.47			0.48		0.49	
B	3.3	3.4	3.5	3.6	3.7	3.8	3.9	4.0			
$f(B)$	0.50			0.51		0.52	0.53				

Table 24

Month	IV	V	VI	VII	VIII	IX	X
$t_s - t_a$ [°C]	2.9	3.8	3.9	3.0	3.4	2.7	1.6

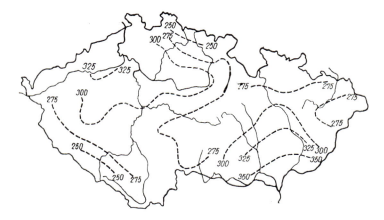

Fig. 4.18. Isolines of evaporation from free water surface [mm] in Czechoslovakia

For a general estimate of evaporation from a water surface, a map of isolines of average evaporation from water surfaces in June, July, and August was constructed *Fig. 4.18*. The average monthly maximum evaporation corresponds to about 40 per cent of the value indicated by the isolines, evaporation during the whole vegetation season to 185 per cent and average annual evaporation to 220 per cent of values indicated by the isolines.

Problem 15

Calculate the approximate values of average daily and monthly evaporation from free water surface during the vegetation season in basin *P* in which ponds are to be built. Long-term averages obtained from the nearest meteorological station are given in *Table 25*.

Table 25

Month	Average air temperature [°C]	Average relative humidity [per cent]	Average velocity [°B]
IV	7.9	74	1.9
V	12.0	73	1.6
VI	15.8	73	1.6
VII	17.0	77	1.5
VIII	16.6	78	1.5
IX	13.1	79	1.4
X	8.4	85	1.5

Equation (4.48) will be used month by month. The values of e_s and e_a will be computed first. For July, for instance, in *appendix 6*, for air temperature of 17.0 °C, the tension of saturated water vapour is indicated as 14.53 mm Hg. For the relative air humidity of 77 per cent, the tension of the water vapour will be $14.53 = 0.77$, or 11.19 mm Hg. *Table 24* will be used to obtain water temperature from air temperature. For July, the difference between the air and the water temperature is 3.0 °C. The water temperature will be $17.0 + 3.0 = 20.0$ °C. In *appendix 6 (page 310)*, the tension of saturated water vapour indicated for this temperature is 17.54 mm Hg.

The value of f (B) for July is, according to *Table 23*, 0.42 (for 1.5 °B). The values of $e_s = 17.54$, $e_a = 11.19$ and f (B) = 0.42 are substituted in eq. (4.44):

$$(17.54 - 11.19)\, 0.42 = 2.7 \text{ mm.}$$

The average daily evaporation in July is thus 2.7 mm. The average monthly value $2.7 \times 31 = 84$ mm. For other months, the calculations are indicated in *Table 25*.

The approximate evaporation from the water surface during vegetation season in the basin is 479 mm.

Evaporation from free water surface is often advantageously computed by converting the pan evaporation data into lake (reservoir) evaporation.

Such methods for conversion of measured pan evaporation into lake (reservoir) evaporation and other formulae based on energy budget and aerodynamical effects may be found in the WMO Technical Notes Nos. 83 and 97, as well as in the WMO Guide to Hydrometeorological Practices (Section A4).

Evaporation from soil and plants (evapotranspiration)

The evapotranspiration may be either potential or actual (*see page 19*). The computation of potential evapotranspiration is most important for irrigation practice, where it corresponds to the net irrigation water requirement. Among the methods used by irrigation hydrologists for this purpose, the Blaney—Criddle (in the USA) and the Alpatyev (in the USSR) approaches are the most representative. They are based on the equation

$$E_{TP} = KF = \Sigma kf, \qquad (4.49)$$

where E_{TP} = potential evapotranspiration for a given period,
$\quad\quad k$ = experimental coefficient of water requirement, different for different plants and months of the growing season,

$$f = pt,$$

where p = percentage of sunshine from the annual total,
$\quad\quad t$ = average air temperature [°C].

To obtain values of evapotranspiration from free water evaporation data, Penman recommends multiplying the values obtained by his formula by factors ranging from 0.6 in winter to 0.8 in summer.

A more accurate computation of potential evapotranspiration for irrigation purposes allow the so-called biological curves of water requirement, which express the ratio of potential evapotranspiration to free water evaporation as it varies during the growing season, with a different probability of occurence. Such curves have been developed in Czechoslovakia by the Irrigation Research Institute (Dvořák, Pýcha 1962).

Actual evapotranspiration

While the lysimetric measurements (*see page 62*) give a certain indication of point values of actual evapotranspiration, the long-term average areal values for a basin may be computed only by the basin water balance, as the difference of measured precipitation and runoff. For smaller areas and very small basins, however, the actual evapotranspiration may be ascertained by the soil water balance method consisting of a daily evaluation of the moisture stored in the soil.

The following equation is used:

$$hR_2 = hR_1 - hE_e + hP - hF_d, \qquad (4.50)$$

where hR_2 = the moisture content in the soil profile typical for a given region at the end of the day [mm],

hR_1 = the moisture content at the beginning of the day [mm],

hE_e = actual daily evapotranspiration [mm/day],

hP = daily precipitation [mm],

hF_d = seepage (flow) into the groundwater [mm].

The seepage into groundwater occurs only if $(hR_1 - hE_e + hP)$ is greater than the field water capacity of the soil.

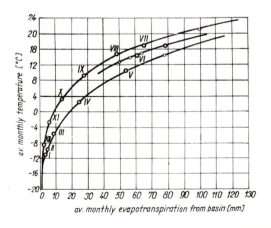

Fig. 4.19. Evapotranspiration from basin (Kuzin)

With this method, the above-mentioned biological curves of water requirement may be derived. The moisture content in the soil is measured either by the classical weighing method or by nuclear probe. In basins containing various types of vegetation (forests, field crops and pastures) and different soils, evapotranspiration can be estimated, in the absence of observations of elements of water balance, with the help of empirical graphs.

A. I. Kuzin's graph (*Fig. 4.19*) is used in the USSR for a very approximate estimate of monthly average evapotranspiration of a basin, from the average monthly temperature. This graph was devised for the climatic conditions in the Upper Volga basin.

4.6   Discharge computations from measured data

(a)   Direct measuring (velocity-area method)

Occasional direct measurements of stream flow by the methods described in chapter 2 on *page 101* and continuous measurements of water stages provide the most common way of streamflow gauging on sites with stable channel control. The rating curve remains relatively stable and may be fitted from measured points, either by inspection or by using graphical or analytical methods for curve fitting as indicated in *problems 3, 4* and *5* on *pages 134 – 147*.

In fitting the rating curve, it must be kept in mind, however, that function $Q = f(H)$ is only an approximation of the complete function indicated by the Chézy equation, for instance, in which $Q = f(H, A, I, n)$. Hence, the discharge is a function of the water stage H, the cross-section area A (expressed by the hydraulic radius R or depth of water in channel h), hydraulic slope I and the roughness of the channel n (after Manning). The fitting of a single curve $Q = f(H)$ assumes that A, I, and n remain constant in time for the given water stage and discharge. This condition, however, is very often not met. Variations of the cross-section may be ascertained rather easily during average discharges. But during floods, and particularly at peaks, the channel is usually eroded and, as the flood recedes, scours are filled again. Hence, unless the control is entirely stable, errors are caused by cross-section variations.

The hydraulic slope and its measurement are sources of additional error. It has been ascertained on many streams, at sites with apparently regular surfaces, that the slope varies considerably in space. The experiments of the Prague Hydraulic Research Institute indicated, however, that the reading of several stage gauges is sufficiently precise for measur-

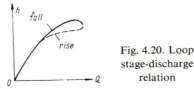

Fig. 4.20. Loop
stage-discharge
relation

ing of the slope (*see page 117*). This is necessary, particularly during floods. As the flood wave rises, the slope increases; it passes a maximum and declines as the flood recedes. Thus, two different water stages in the upper part of the rating curve may correspond to the same flow and vice versa (*Fig. 4.20*). The rating curve forms a loop, called hysteresis stage-discharge relation.

The effect of the roughness of the channel (expressed by Manning as n) is very important. A considerable fluctuation in roughness exists in smaller streams, where n changes with the season as a result of the process of growth of aquatic vegetation.

It is therefore evident that, in plotting a stage-discharge relation, an excessive precision is almost superfluous. A simple fitting as indicated in *problem 3* is often entirely sufficient.

To check a stage-discharge relation, it is recommended to plot the relations

$$v = f_1(H)$$

and

$$A = f_2(H),$$

where v = average cross-sectional velocity, A and H as above. Only if the plotted points form more or less regular curves, a more accurate fitting of the stage discharge relation is advised.

(b) Indirect measurements (slope-area method)

Often it is necessary to estimate a discharge which has not been measured directly; for instance, in extrapolating the rating curve above the measured discharge, in calculating flood peak discharges from marks left in the flood plain, etc.

In all these cases, discharge Q may be computed from the cross-sectional velocity v_{mean} from the so-called open-channel flow formulae. The *Chézy equation* is the one used most often:

$$v_{mean} = c \sqrt{RI}, \tag{4.51}$$

where c = the Chézy coefficient,

$\quad R$ = hydraulic radius = $\dfrac{A}{W}$,

$\quad A$ = cross-sectional area,

$\quad W$ = wetted perimeter,

$\quad I$ = hydraulic slope of water surface (energy gradient) in absolute units.

If v_{mean} is in m/s, A and W must also be in metres.

After calculating the velocity, the discharge Q is computed from the equation

$$Q = A v_{mean}. \tag{4.52}$$

There are several open-channel flow formulae — those of *Manning, Strickler*, etc. (see Chow, 1959). For open channels, eq. (4.51) is satisfactory, providing that the velocity factor c has been properly assumed. According to measurements made on many rivers in Czechoslovakia, the following Chézy coefficients are satisfactory from theoretical as well as practical standpoints.

Ganguillet — Kutter

$$c = \frac{23 + \dfrac{0.001\,55}{I} + \dfrac{1}{n}}{1 + \left(23 + \dfrac{0.001\,55}{I}\right)\dfrac{n}{\sqrt{R}}}. \tag{4.53}$$

Pavlovsky

$$c = \frac{1}{n} R^y,   \tag{4.54}$$

where

$$y = 2 \times 5\sqrt{n} - 0.13 - 0.75\sqrt{R}(\sqrt{n} - 0.10)   \tag{4.55}$$

or, in simpler form,

$$\begin{aligned} y &= 1.5\sqrt{n} \quad \text{for} \quad R < 1, \\ y &= 1.3\sqrt{n} \quad \text{for} \quad R > 1. \end{aligned} \tag{4.56}$$

The choice of Manning roughness n is the crucial step in the computation of c. If it is possible to measure the relative roughness or the average value of channel resistance $\dfrac{R}{d_{50}}$, where d_{50} is the average dimension of the material which forms the channel bed (gravel or sand) corresponding to the 50 per cent value of the mass curve of bed material sample (sieve with apertures 0.2, 0.5, 1, 2, 3, and 5 cm), the following equations may be used (originated by the Prague Hydraulic Research Institute):

$$c = 17.7 \log \frac{R}{d_{50}} + 13.6,   \tag{4.57}$$

or simply

$$c = 18 \log \frac{R}{d} + 13.   \tag{4.58}$$

Table 26 (according to Sribny) is used frequently in Czechoslovakia for the selection of the roughness n.

Table 27 may be used to determine the exponent y for eq. (4.54).

Table 28 gives the Chézy coefficient c (according to Pavlovsky) for $0.05 \text{ m} < R < 5 \text{ m}$ and $0.11 < n < 0.040$, and may be readily used for slope-area discharge computations.

In *Table 28*, it is not necessary to interpolate, since the closest suitable number is still within the margin of error, committed by the selection of n.

The use of the slope area method for extrapolation of stage-discharge relation and for calculating peak discharge is illustrated by the following problems:

Table 26

Number	Type of channel	n	$\dfrac{1}{n}$	γ
1	Natural channel in exceptionally good condition, free flow	0.025	40	1.25
2	Channel of a stable character, large and medium rivers, good conditions	0.033	30	2.00
3	Free flow in channel, minor unevenness of the bed or minor meanders, channel without aquatic vegetation	0.040	25	2.75
4	Channel of large or medium sized rivers with considerable aquatic vegetation and deposits, meanders. Streams with a large amount of gravel bed load	0.050	20	3.75
5	Channel of a perennial stream, clogged, flood plain with trees, brooks with very rough bed	0.067	15	5.50
6	Rivers and flood plains with slow flow, deep pools, swift brooks with broken warped surfaces	0.080	12.5	7.00
7	Flood plains or rivers as in No. 6, but heavily meandering. Wild brooks with waterfalls which drown the human voice	0.100	10	9.00
8	Streams in marshes, in places stagnant. Forested flood plains	0.133	7.5	12.00
9	River bed completely clogged with stones, very low slope	0.200	5	20.00

Note: n = Manning roughness,
γ = Bazin roughness.

Table 27

$\dfrac{1}{n}$	$\geqq 100$	70	55	40	25	12.5	5
y	$\dfrac{1}{8}$	$\dfrac{1}{7}$	$\dfrac{1}{6}$	$\dfrac{1}{5}$	$\dfrac{1}{4}$	$\dfrac{1}{3}$	$\dfrac{1}{2}$

Table 28

Hydraulic radius [m]	Manning n							
	0.011	0.013	0.014	0.020	0.025	0.030	0.035	0.040
	Chézy coefficient c							
0.05	61.3	48.4	33.2	26.1	18.6	13.9	10.9	8.7
0.06	62.8	50.1	34.4	27.2	19.5	14.4	11.5	9.3
0.07	64.1	51.3	35.5	28.2	20.4	15.5	12.2	9.9
0.08	52.2	52.4	36.4	29.0	21.1	16.1	12.8	10.3
0.10	67.2	54.3	38.1	30.6	22.4	17.3	13.8	11.2
0.12	68.8	55.8	39.5	32.6	23.5	18.3	14.7	12.1
0.14	70.3	57.2	40.4	33.0	24.5	19.1	15.4	12.8
0.16	71.5	58.4	41.8	34.0	25.4	19.9	16.1	13.4
0.18	72.6	59.5	42.7	34.8	26.2	20.6	16.8	17.0
0.20	73.7	60.4	43.6	35.7	26.9	21.3	17.4	17.5
0.22	74.6	61.3	44.4	36.4	27.6	21.9	17.9	15.0
0.24	75.5	62.1	45.2	37.1	28.3	22.5	18.5	15.0
0.26	76.3	62.9	45.9	37.8	28.8	23.0	18.9	16.0
0.28	77.0	63.9	46.5	38.4	29.4	23.5	19.4	16.4
0.30	77.7	64.3	47.2	39.0	29.9	24.0	19.9	16.8
0.35	79.3	65.8	48.6	40.3	31.1	25.1	20.9	17.8
0.40	80.8	67.1	49.8	41.5	32.2	26.0	21.8	18.6
0.45	82.0	68.4	50.9	42.5	33.1	26.9	22.6	19.4
0.50	83.1	69.4	51.9	43.5	34.0	27.8	23.4	20.1
0.55	84.1	70.4	52.8	44.4	34.8	28.5	24.0	20.7
0.60	85.3	71.4	53.7	45.2	35.5	29.2	24.7	21.3
0.65	86.0	72.2	54.5	45.9	36.2	29.8	25.3	21.9
0.70	86.8	73.0	55.2	46.6	36.9	30.4	25.8	22.4
0.80	88.3	74.5	56.2	47.9	38.0	31.5	26.8	23.4
0.90	89.4	75.5	57.5	48.8	38.9	32.3	27.6	24.1
1.00	90.9	76.9	58.8	50.0	40.0	33.3	28.6	25.0
1.10	92.0	78.0	59.8	50.9	40.9	34.1	29.3	25.4
1.20	93.1	79.0	60.4	51.8	41.6	34.8	30.0	26.3
1.30	94.0	79.9	61.5	52.5	42.3	35.5	30.6	26.9
1.40	95.7	81.5	62.9	53.9	43.6	36.7	31.4	28.0
1.50	97.3	82.9	64.3	55.1	44.7	37.4	32.7	28.9
2.00	99.3	84.8	65.9	56.6	46.0	38.9	33.8	30.0
2.50	102.1	87.3	68.1	58.7	47.9	40.6	35.4	31.5
3.00	104.4	89.4	69.8	60.3	41.9	49.3	36.6	32.5
3.50	106.4	91.1	71.3	61.5	50.3	42.8	37.4	33.3
4.00	108.1	92.6	72.5	62.5	51.2	43.6	38.1	33.9
5.00	111.0	95.1	74.2	64.1	52.4	44.6	38.9	34.6

Problem 16

Discharges have been measured in a regular, wide, and relatively deep channel on a small stream at water stages varying from 1.3 to 1.5 m. Extrapolate the rating curve to a water stage of 1.8 m, assuming that it was possible to measure *in situ* water depth h and the cross-sectional area A up to this water stage (*Table 29a*).

Table 29a

H [m]	1.3	1.4	1.5	1.6	1.7	1.8
h [dm]	1.3	2.0	2.7	3.3	4.0	4.7
A [dm^2]	377	692	1067	1430	1956	2450
Q [l/s]	280	680	1200	—	—	—

The problem involves an extrapolation of the stage-discharge relation. The Chézy eq. (4.51) and eq. (4.52) will be used.

For a wide channel, the average depth h (hydraulic mean depth) may be substituted to the hydraulic radius R, hence

$$Q = c \sqrt{I} \cdot A \cdot \sqrt{h}.$$

For a deeper stream, it may be assumed that $c \sqrt{I}$ will not rapidly change and the equation may be written:

$$Q = mA \sqrt{h},$$

where $m = c \sqrt{I}$.

For further computation, it is assumed that a linear regression exists between Q and $mA \sqrt{h}$. If plotted in normal rectangular co-ordinates, the maximum observed water stage may be extrapolated by 30 per cent. For this purpose, the values of h are used; h may be computed from the ratio $\dfrac{A}{w}$, where $w =$ the width of the stream at water surface level and $A =$ area of cross-section.

The computations of A, h, and $m = \dfrac{Q}{A \sqrt{h}}$ are in *Table 29b*.

Table 29b

H	1.3	1.4	1.5	1.6	1.7	1.8
$A\sqrt{h}$	430	980	1750	2603	3912	5317
m	0.65	0.69	0.69	—	—	—
Q [l/s]	280	680	1200	1770	2660	3616

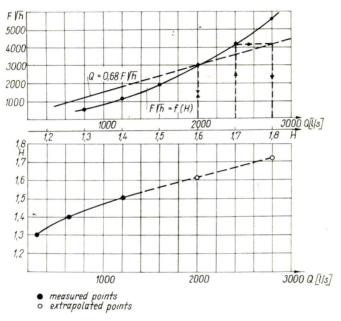

● measured points
○ extrapolated points

Fig. 4.21. Extension of stage-discharge relation

The arithmetic average of m is 0.68; hence, $Q = 0.68F\sqrt{h}$. From this equation, the discharges corresponding to water stages 1.6, 1.7, and 1.8 are either calculated or estimated from the graph in *Fig. 4.21.* For $H = 1.6$, $Q = 1770$ l/s; for $H = 1.7$, $Q = 2660$ l/s; for $H = 1.8$, $Q = 3616$ l/s.

It must be kept in mind that the extrapolated stage-discharge relations are only approximate and that this method may be used only if both

the shape of the channel cross-section and its roughness remain approximately the same during floods and the site is situated in a straight
reach with a constant slope.

Problem 17

A large flash flood occurred on a small stream at night. The water
surface slope was later levelled in a straight reach about 100 m long,

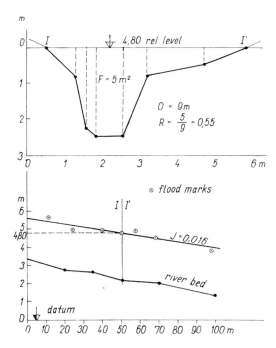

Fig. 4.22. Discharge computation by using area-slope method

according to marks left by the flood on the banks. The areas of cross-
section at the sites of marks were carefully surveyed. The profile of the
water surface was plotted in *Fig. 4.22* ($I - I'$). Calculate the maximum
(peak) flood discharge in this stream reach.

In *Fig. 4.22* data are plotted for the water surface slope and cross-
section area computation. The slope line is fitted by a straight line; the

fitting is done by inspection, in view of the approximation of the whole calculation. The value of the slope is $0.016 = 16$ per cent. The area of the cross-section is $A = 5$ m^2 and the wetted perimeter $W = 9$ m.

The Chézy equation is used with a velocity factor according to Pavlovsky. From *Table 26*, $n = 0.080$, $\dfrac{1}{n} = 12.5$. According to *Table 27*, $y = \dfrac{1}{3}$. The hydraulic radius $R = \dfrac{A}{W} = \dfrac{5}{9} = 0.55$.

Substituting in eq. (4.54)

$$c = 12.5 \times 0.55^{1/3} = 10.25,$$

and then in eq. (4.51)

$$v = 10.25 \sqrt{0.55 \times 0.016} = 10.25 \times 0.094 = 1.0 \text{ m/s}$$

and eventually in eq. (4.52)

$$Q = 5 \times 1 = 5 \text{ m}^3/\text{s}.$$

If the value of n is difficult to ascertain, two extreme values are selected and the computed discharges serve as boundaries of the confidence interval for further consideration.

The stage-discharge relation for a stream which is generally well surveyed and on which there are several gauging stations can be extended by computation of the increase of the peak flood flow along the stream, providing there are no large reservoirs in the basin. The procedure assumes that the peak discharge of a flood has been measured at all gauging stations on the stream. The following relation may then be developed:

$$Q = aA^n, \tag{4.59}$$

where Q = the peak flood discharge measured at a gauging station,

A = area of basin for this gauging station [km^2],

a, n = parameters which are ascertained for the entire basin.

If the parameters of eq. (4.59) can be computed from measured data, the ungauged discharge at a stream-gauging station at which the stage-discharge relation is to be extended may be ascertained by it. The corresponding stage to this discharge has either been recorded or may be established approximately according to marks left by the flood.

Problem 18

An exceptionally high flood occurred on stream L on 17 and 18 July. The peak discharge HQ was not measured at the station in the upper reaches of the stream (catchment area of station 200 km²). Discharges were, however, measured at gauging stations B, C, D, and E further downstream. Determine the corresponding discharge at station A.

The data are plotted in a log-log (double logarithmic) paper (*Fig. 4.23*). A straight line may be fitted to them by inspection and extended. According to the graph for $A =$ 200 km², $Q = 105$ m³/s. Parameters of eq. (4.59) may also be calculated.

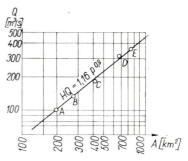

Fig. 4.23. Extension of rating curve

$$n \cong \frac{\log Q_E - \log Q_B}{\log A_B - \log A_B}, \qquad (4.60)$$

$$a = \frac{Q_E}{A_E^n} = \frac{Q_B}{A_B^n}. \qquad (4.61)$$

Data:

Station	B	C	D	E
Catchment area [km²]	273	414	656	801
HQ [m³/s]	132	180	312	356

The following numerical values are substituted:

$$Q_E = 356 \text{ m}^3/\text{s}, \qquad Q_B = 132 \text{ m}^3/\text{s},$$
$$A_E = 801 \text{ km}^2, \qquad A_B = 273 \text{ m}^3/\text{s}.$$

Equations (4.60) and (4.61) will then read:

$$n = \frac{\log 356 - \log 132}{\log 801 - \log 273} = \frac{0.430}{0.467} \cong 0.90,$$

$$a = \frac{356}{801^{0.9}} = \frac{132}{273^{0.9}} = 0.87,$$

Equation (4.59) will be:

$$Q = 0.87A^{0.9}.$$

If $A = 200$ km^2 is substituted in the above equation

$$Q = 0.87 \times 200^{0.9} = 1.16 \times 1178 = 102 \text{ m}^3/\text{s}.$$

The result is approximately the same as the graphical solution. This method can also be used for interpolating the stage-discharge relation between two observed water stages and discharges. Results are naturally more reliable for interpolation than for extension. For interpolation, the straight line can be fitted in log-log co-ordinates by one of the methods given on *pages 135 – 147*. For extension, it is not necessary to use precise fitting methods, in view of the relative inaccuracy of the results.

4.7   Mean annual runoff (yield), long-term average annual runoff, and the distribution of runoff during the year

The mean annual runoff for a given year $Q(y)$ is the arithmetic average of the mean daily $Q(d)$ or mean monthly flows $Q(m)$ for this year.

For a gauged basin with a sufficient network of stream-flow gauging stations, the data are readily available, either in the publication of the appropriate responsible agency (hydrological service) or may be supplied on request by this agency. The World Meteorological Organization has recently published a list of the agencies in charge of hydrological data collection throughout the world. Addresses of depositories of data are also included in this publication (Reports on WMO/IHD Projects, Report No. 10: 'Organization of Hydrometeorological and Hydrological Services', WMO Secretariat, Geneva, Switzerland, 1969).

In Great Britain, these data are published in the Surface Water Yearbook of Great Britain published by HMSO. In all central and eastern European countries, similar yearbooks are published on an annual basis.

In Czechoslovakia, the Hydrometeorological Institute (combining the meteorological and hydrological services) supplies on request data on runoff from basins which are not directly gauged. Such data are extrapolated by hydrological analysis and correlation techniques from gauged basins.

Some of the simplest of these techniques are indicated below. The most general picture and source of preliminary data may be a *map of the long-term average annual runoff*. Such a map for Czechoslovakia is shown in *Fig. 4.23a*. For the preparation of such maps, direct stream-gauging data are used, but they are usually supplemented by data included in precipitation and/or evaporation maps. A detailed discussion of the preparation of co-ordinated runoff, precipitation, and evaporation maps is contained in a report by T. J. Nordenson, published by the World Meteorological Organization (Reports on WMO/IHD Projects, Report No. 6: 'Preparation of co-ordinated precipitation runoff and evaporation maps,' WMO Secretariat, Geneva, Switzerland, 1968).

Long-term annual average runoff data taken from a map may, however, serve only general planning purposes. Design data are to be computed more exactly, and the following simple procedure may be used for basins where direct gauging data are lacking.

This simple method used by the Hydrometeorological Institute in Czechoslovakia is based on the assumption of a linear regression between annual precipitation and runoff, as illustrated in *Fig. 4.1* for the Elbe basin. The equation of straight-line regression is

$$hQ = ahP + b, \tag{4.62}$$

where hQ = annual runoff [mm],

$\quad hP$ = annual precipitation [mm],

$\quad a, b$ = basin parameters.

If the values of hQ and corresponding hP are known for similar basins, a and b can be calculated by methods indicated in chapter 3 (*page 135*).

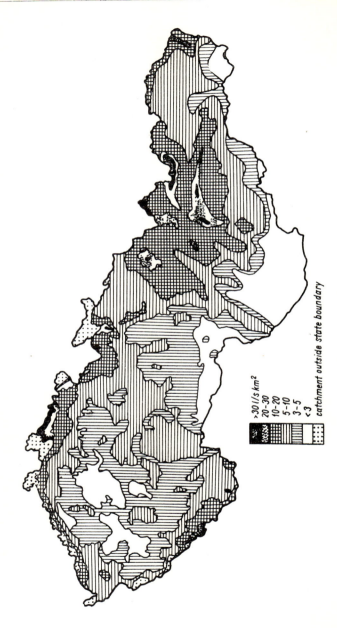

>30 l/s km²
20–30
10–20
5–10
3–5
<3
catchment outside state boundary

Fig. 4.23a. Isolines of runoff in Czechoslovakia

Problem 19

The following values were ascertained for annual average precipitation and runoff in two similar basins of a region.

hP	hQ
623 mm	119 mm
796 mm	338 mm

These values, inserted into eq. (4.62), yielded the values of $a = 1.26$ and $b = 666.0$.

Thus, eq. (4.62) for the given hydrological region is

$$hQ(y) = 1.26hP(y) - 666.0.$$

This method is mostly used for sub-basins of a larger basin.

(a) Distribution of runoff in time

The long-term average annual runoff indicates the total yield of water from the basin but does not provide any insight into how this yield is distributed in time. Such a distribution may be ascertained by characteristics of:

(a) variations of annual runoff from year to year;

(b) variations of runoff within a particular year or in average year, either by months or by seasons;

(c) variations of daily discharges throughout one particular year or in average year.

All those characteristics may be computed from stream-gauging data as indicated in section 4.3. In the absence of such data, several methods of hydrological analysis may be used and some are indicated below.

Variation of mean annual runoff from year to year

(a) From a series of mean annual discharges $Q(y)$ provided by the agency in charge of stream-gauging, a coefficient k_i for each year may be calculated from the equation

$$k_i = \frac{Q(y)_i}{MQ(y)}. \tag{4.63}$$

These coefficients indicate whether the year's water yield was larger or lower with respect to the long-term average. For economic consideration of design data, however, the frequency of occurrence is most important.

If the series of observations is for n years and if it is necessary to ascertain $Q(y)$ with a return interval of N years, and $N = n \pm 2$ eq. (4.20) may be used to calculate the probability p and return interval T:

$$p \text{ per cent} = \frac{m - 0.3}{n + 0.4} \, 100; \qquad T = \frac{1}{p} \qquad \text{or} \qquad T = \frac{1}{1 - p}.$$

The procedure is indicated in *problem 7*, where annual precipitation may be replaced by annual runoff.

If the required return interval is much larger than the period of observation, a theoretical frequency distribution curve may be fitted to the series. The Pearson Type III curve is used in Czechoslovakia for this purpose.

The procedure is described in *problem 8*.

(b) In basins which have no stream gauging data for $Q(r)$, C_v and C_s may be calculated by assuming an empirical correlation between C_v of a series of mean annual discharge, and $MQ(r)$, that is, the long-term average annual discharge and the area of basin A. Such correlation has been ascertained and is considered as valid on many basins in the USSR. In Czechoslovakia, the Hydrologic Institute of the Slovak Academy of Sciences (Svoboda 1965) has demonstrated the existence of such correlation for almost all basins of the country, which were divided into four groups for this purpose. This empirical correlation is a physical phenomenon and may thus be computed for other regions or countries, including the UK. The correlation has the following general form:

$$C_v = \frac{a}{q^n} + b \log \frac{c}{A}, \tag{4.64}$$

where $C_v = \dfrac{\sigma}{\bar{x}} =$ coefficient of variation of annual runoff values,

$q = \dfrac{MQ(y)}{A} =$ long-term annual average runoff per unit of basin area [l/s km],

$A =$ basin area [km²],

$a, b, c, n =$ parameters computed from gauged catchments and valid for larger hydrological regions (groups of catchments).

If C_v is computed by the above method, which assumes that $MQ(y) = \bar{x}$ is known, C_s may be safely assumed as $C_s = 2C_v$, in view of the approximative character of this method. Knowing $\bar{x}$, C_v, and C_s, the theoretical frequency curve of $Q(y)$ may be plotted and/or its particular points computed by the procedure indicated in *problem 8*, or by any other suitable method.

Problem 20

A reservoir is to be designed on an ungauged small river with a catchment area A of 13.9 km², located in a hydrologic region with a long-term average annual runoff per unit of catchment area $q = 1.39$ l/s km². Compute the annual runoff value of 80 per cent probability (return interval $T = 5$ years) needed for the design of the volume of the reservoir.

The parameters of eq. (4.64) for the above hydrological region are:

$$a = 0.793; \qquad b = 0.0782; \qquad c = 811; \qquad n = 0.447.$$

By substituting these parameters in eq. (4.64), the value of C_v is computed:

$$C_v = \frac{0.793}{1.399 \cdot 0.447} + 0.0782 \log \frac{811}{13.9} = 0.82.$$

The coefficient of assymetry $C_s = 2C_v = 1.64$.

The table of values of Pearson Type III frequency curve in appendix 5 indicates for $C_s = 1.64$ and $C_s = 2C_v$ the frequency factor K for $p = 80$ per cent:

$$K_{80\%} = 0.352.$$

Subsequently

$$q(y)_{80\%} = Mq(y) \, K_{80\%} = 1.39 \times 0.352 = 0.489 \text{ l/s km}^2,$$

and

$$Q(y)_{80\%} = q(y)_{80\%} \, A = 0.489 \times 13.9 = 6.8 \text{ l/s}.$$

The average annual runoff yield of 6.8 l/s has a probability of 80%. The runoff in the basin is therefore larger than 6.8 l/s in 80 out of 100 years, or, on average lower once in 5 years than 6.9 l/s. This runoff is therefore relatively very low.

Variation of runoff during the year

The variation of runoff during the year is best expressed by monthly flow $Q(m)$ or long-term average monthly runoff $MQ(m)$. On gauged streams, the necessary data are readily available.

The monthly runoff may also be computed with respect to its probability of occurrence if the series of mean monthly discharges $Q(m)$ are used for such frequency analysis. For a short return interval, the procedure of *problem 7* may be used and, for extrapolation, a theoretical frequency distribution may be fitted as in *problem 8*.

Table 30

Section	Subsection	XI [%]	XII [%]	I [%]	II [%]	III [%]	IV [%]	V [%]	VI [%]	VII [%]	VIII [%]	IX [%]	X [%]	I-XII [%]
										Month				
I	—	7.5	7.5	8.5	12.5	15	14	7.5	5	6.5	5	4.5	6.5	100
II	—	6.5	6	7.5	11	16	15	11	8.5	7.5	4	3	4	100
III	a	5.5	6	5	6.5	9	15	19	8	8.5	5	6	6.5	100
	b	7.5	6.5	6	6.5	9.5	22.5	16	5	4	4.5	5.5	6.5	100
IV	a	7	8.5	7	11	12	10	7	6	8.5	6.5	7.5	9	100
	b	7	8.5	7	11	12	10	10	5.5	8.5	7	6	7.5	100
	c	7	8.5	7	11	12	10	9	9	10	6	4.5	6	100
V	—	7.5	7	7.5	8	10	9	8.5	8	8.0	9	9	9	100

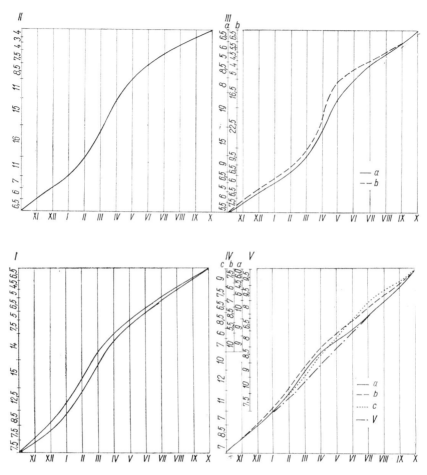

Fig. 4.24. Mass curves of monthly flows

On small streams, without a gauging station, the procedures of extra-polation from similar gauged basins must be used.

For preliminary studies, typical mass curves of long-term average monthly runoff may be plotted for groups of basins in relatively homogeneous hydrological regions. Four such regions have been ascert-ained and corresponding curves plotted for the entire basin of the Elbe

(Labe) in Czechoslovakia by the author. They are shown on *Fig. 4.24.* The curves are based on runoff distribution into months, expressed in percentages of annual runoff. It is to be noted that for groups III and IV, subgroups have been ascertained comprising smaller catchments within larger hydrological regions. Two such subgroups were derived for region III and three for region IV.

In this connection, it should be pointed out that minimum monthly runoff, that is, exceptionally dry months, need not necessarily occur in the same year that has an extremely low yearly runoff. Runoff in critical dry periods is discussed later in connection with the computation of low flows.

Table 31

	Seconds
1 day	86 400
1 ordinary year	31 536 000
1 leap year	31 622 400
1 average year	31 557 600
1 month (30 days)	2 592 000
1 month (31 days)	2 678 400
February (28 days)	2 419 000
February (29 days)	2 506 000
average February	2 440 600

Data in *Table 31* facilitate the conversion of discharge [m³/s] to volume of runoff [m³] and vice versa.

Problem 21

Estimate the average monthly runoff distribution of the average annual runoff $MQ(y)$ from the basin in *problem 20*.

In *problem 20*, $Mq(y)$ was 1.39 l/s km², and

$$MQ(y) = Mq(y) \, A = 1.39 \times 13.9 = 19.3 \text{ l/s}.$$

The average annual volume of runoff is computed with the help of *Table 31*:

$$0.0199 \text{ m}^3/\text{s} \times 31\ 557\ 600 \text{ s} = 609\ 052 \text{ m}^3.$$

The basin is situated in the region of group I (*see Fig. 4.24*). So, for example, in the month of November, the monthly runoff will be 7.5 per cent of the annual runoff

$$\frac{609\,062 \times 7.5}{100} = 45\,680 \text{ m}^3.$$

The average monthly discharge in November will be (*see Table 31*).

$$\frac{45\,680}{2592} = 17.6 \text{ l/s} = 0.0176 \text{ m}^3/\text{s}.$$

The computations are indicated in *Table 32*.

Table 32

Month	Volume [m³]	Per cent	Time [s × 10³]	Rate of flow, n [l/s]	Notes	Approximate rate of flow [l/s]
XI	45 680	7.5	2 592	17.6		17.3
XII	45 680	7.5	2 678.4	17.1		17.3
I	51 770	8.5	2 678.4	19.3		19.7
II	76 133	12.5	2 440.6	31.1		29.0
III	91 309	15.0	2 678.4	34.1	maximum	34.7
IV	85 269	14.0	2 592	22.9		32.4
V	45 680	7.5	2 678.4	17.1		17.3
VI	30 453	5	2 592	11.7		11.6
VII	39 589	6.5	2 678.4	14.8		15.1
VIII	30 453	5	2 678.4	11.4		11.6
IX	27 408	4.5	2 592	10.6	minimum	10.4
X	39 589	6.5	2 678	14.8		15.1
						Σ 231.5
I—XII	609 063	100	31 557.6	—	$MQ(y) = \dfrac{231.5}{12} = 19.3$	

In view of the physical approximation of the monthly percentages of the annual runoff, the calculation may be simplified by assuming all

the months of the year to be of the same time duration. It then follows
that

$$MQ(m) = \frac{m\% \, MQ(y)}{\frac{100}{12}} = \frac{12m\% \, MQ(y)}{100}, \qquad (4.65)$$

where $m\%$ is the percentage share of each individual month in the total
annual runoff volume. Thus, for November,

$$MQ_{Nov} = \frac{7.5 \times 19.3 \times 12}{100} = 17.3 \, \text{l/s}.$$

The results computed in this simplified way are given in *Table 32* for
purposes of comparison.

Flow duration curve of mean daily discharges

When it is necessary to determine the variations of runoff in a year
with respect to their magnitude and not according to their calendar
sequence, the flow duration curves of mean daily discharge are plotted.
This curve may cover one particular year (as constructed in *problem 6*),
or a series of years, during which the mean daily discharges have been
measured.

In the latter case, the procedure consists in considering the entire
period (n years) as 100 per cent and the calculation is the same as in
problem 6, the only difference being that all the n-time 365 observations
(n = number of years of observation) are distributed into the intervals.
The scale of duration is again 365 days, but the values represent an
average for the whole period.

An alternative method is as follows.

For each year, a flow duration curve is plotted and the values of 10,
25, 50, 75, and 90 per cent of duration are ascertained (in central Europe
these values are called upper and lower deciles, quadrils, and the median).
For each of the corresponding duration percentage, an arithmetic average
is computed and the average curve is plotted from these averages.

As several authors have ascertained in Czechoslovakia, typical flow
duration curves for long-term averages of daily mean discharges (plotted
by one of the methods discussed above) may be abstracted for groups
of basins in homogeneous hydrological regions. In general, these regions

correspond to those ascertained for typical monthly variations of runoff.

The author has abstracted such typical flow duration curves for small basins in regions of the Elbe Basin in Bohemia. It is believed that a similar approach may also be adopted for larger basins in regions with sparse stream-gauging networks, particularly with respect to general preliminary hydrological studies for hydro-power development.

Values of characteristic points of the typical flow duration curves for groups of basins in the Elbe (Labe) catchment reduced to the ratio of long-term average annual flow $\frac{MQ_n}{MQ(y)}$ are given in *Table 33*.

Table 33. Approximate rate of flow exceeding n days (MQ_n)

Region	MQ_n												
	$n = 30$	60	90	120	150	180	210	240	270	300	330	355	364 days
I	1.90	1.48	1.13	0.88	0.71	0.59	0.49	0.41	0.34	0.28	0.20	0.13	0.07
II	2.20	1.48	1.14	0.94	0.79	0.80	0.63	0.53	0.47	0.41	0.33	0.25	0.21
III	1.97	1.41	1.18	1.02	0.90	0.79	0.70	0.62	0.55	0.48	0.41	0.33	0.25
IVa	1.80	1.37	1.13	0.95	0.85	0.76	0.70	0.66	0.62	0.56	0.52	0.45	0.40
IVb	2.25	1.50	1.20	0.98	0.75	0.65	0.50	0.46	0.40	0.33	0.25	0.14	0.07
V	1.34	1.23	1.15	1.09	1.04	0.99	0.94	0.88	0.85	0.79	0.64	0.44	0.23
VI	2.80	1.65	0.98	0.75	0.57	0.43	0.35	0.29	0.23	0.18	0.13	0.07	0.03

Note, in particular, Group VI. This flow duration curve corresponds to very small basins (4 to 10 km² area) in moderately forested low-elevation mountains. Such typical flow duration curves may be abstracted for hydrological regions in different countries. They are, however, closely connected to the hydrological character of the region and cannot be transposed from one region to another.

The typical curves in *Table 33* correspond to years with long-term average annual discharge and the figures given are expressed as a ratio to this discharge. The curves will have a different shape for years with a higher or lower annual flow; curves for dry years will be discussed in connection with minimum runoff.

Problem 22

Plot the flow duration curve of mean daily discharge for a year with long-term average annual runoff $MQ(y)$ for the basin considered in *problem 20*.

The data given in *Table 33* for Group I apply to this basin. The values of characteristic points of the duration curve in *Table 33*, are multiplied by $MQ(y) = 19.3$ l/s (*see Table 34*). From computed values of MQ_n, the duration curve for an average year is plotted (*Fig. 4.25*).

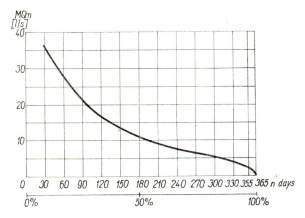

Fig. 4.25. Typical flow duration curve

Table 34

n	30	60	90	120	150	180	210	240	270	300	330	355	364
$\dfrac{MQ_n}{MQ(y)}$	1.9	1.48	1.13	0.88	0.71	0.59	0.49	0.41	0.34	0.28	0.20	0.13	0.07
MQ_n [l/s]	35.9	28.0	21.4	16.6	13.4	11.2	9.3	7.7	6.4	5.3	3.8	2.5	1.32

The duration curve of mean daily water stages is also important. It is particularly necessary for the location of levels of drainage outlets, for flood control studies and navigation purposes.

It can be derived from the typical flow duration curve (if there are no direct observations of water stage), using a suitable stage-discharge relation. This can be calculated, for instance, on the basis of the slope-area method, when projects of new stream channels are considered.

4.8   Flood flow

The flood characteristics and flows are closely connected with the problem of suitable design criteria for structures in water resources projects. These include spillway design and capacity of reservoirs, dimensions of river channels, culverts and bridges, flood control, and storm sewer systems, etc. Most projects being evaluated from the cost-benefit aspect, the design criteria and corresponding data must include parameters for such cost-benefit considerations. It has already been mentioned that the frequency analysis of hydrological phenomena provides these parameters. They become of paramount importance for flood flow estimation. On the other hand, in some cases, the prevention of any risk of failure of a structure prevails over economic considerations. This is particularly true if human lives could be considerably endangered by floods. In such cases, the physical maximum of the phenomenon is sought. For instance, the probable maximum precipitation that may occur is used as input for the synthesis of the flood hydrograph.

Techniques suitable for both approaches are considered below. However, emphasis is given to estimation of flood flow and design data for the purposes of projects in smaller basins. Flood design data for projects on large rivers, involving important capital investments, generally warrant exhaustive and often sophisticated hydrological studies and surveys, using techniques beyond the scope of this book. Highly specialized hydrometeorologists and hydrologists may develop such techniques specifically for the purpose. Even in such cases, however, the principles indicated below constitute the basic approach.

There are many papers and publications referring particularly to this problem. Among the recent ones is the Technical Note No. 98 on 'Estimation of Maximum Floods', published by the World Meteorological Organisation (WMO Publication No. 233-TP. 126, WMO Secretariat, Geneva, Switzerland, 1969). Both approaches mentioned above are included in this publication.

In the USSR and, to a certain extent, in Czechoslovakia and some other countries of central Europe, a standard classification of projects with respect to the frequency (recurrence interval N) of a flood peak for which the project will be designed is based on the capital investment in the structure.

For structures in smaller basins, such classification is indicated in *Table 35*.

Table 35

Structure	Design flood peak of a recurrence interval of N years
Spillways of small reservoir dams in the countryside, not endangering urban residences or other structures in case of failure	10—20
As above but located so as to endanger other structures or urban residences in case of failure	50—100
Channel improvements on small streams in open country, where flooding is not desirable	3—5
Channel improvements on small streams in areas threatening safety of buildings and urban residences	50—100
Culverts and small bridges on main highways	50—100
Culverts and small bridges on less important highways and roads	30—50

From this table, it is evident that a suitable method must be used to ascertain floods of a recurrence interval of N years for $N = 3, 10, 20, 30, 50,$ and 100 years.

It should be emphasized once more that a hundred-year flood occurs once in 100 years in a long-term average — for instance, ten times in 1000 years. It can, however, occur several times in successive years, and certainly an even larger flood may occur in this period. For example, for structures dimensioned for HQ_{100} which have a life span of 100 years, there is a 60 per cent probability, that, during the 100 years of their existence, they will have to accommodate discharges one to three times larger than HQ_{100}. This must be duly taken into account in the design of such structures.

The selection of the method for the computation of the flood flow is, however, also dependent on the raw data available. As already mentioned on *page 129*, three basic alternatives may occur.

(a) At the site for which the flood flow is to be computed, a stream-gauging station is located, with records of flood flows covering a relatively long period.

(b) No stream-gauging stations are located at the site or in its vicinity and no flood flows have been gauged in the basin.

(c) Some flood flows have been gauged directly at the site under consideration or in its vicinity, but the records cover only a short period. Alternatively, long-term records of flood flow exist for a gauging station in the basin, usually downstream from the site.

Although, in alternative (a) both frequency analysis of flow records and synthesis of an extrapolated maximum flood hydrograph may be computed, in alternatives (b) and (c) a synthesis of the hydrograph must be attempted on the basis of available hydrometeorological data and other factors of runoff. The hydrometeorological data may be processed by frequency analysis, and thus the frequency factor introduced into the computation.

In all alternatives, with the exception of frequency analysis of stream flow data, the *synthesis of the flood hydrograph* or of its elements is the basic component of the computations.

For this reason, an analysis of the elements of a flood hydrograph is given below.

The so-called single-peak flood wave (*Fig. 4.26*) is the most common form of flood hydrograph. A double-peak hydrograph (*Fig. 4.26*) may be generated by consecutive storms. Complex hydrographs from multiple storms are also frequent.

In analysing a hydrograph, its origin in time must be determined (*Fig. 4.26, point A*). On small basins, it closely approximates the time of the beginning of the rainfall or immediately after it.

The part of a hydrograph preceding the origin A, if it follows a period without rain, may be considered as the so-called *recession (depletion) curve* of the basin. It indicates the depletion of the storage of water in the basin in a period without rain, hence without direct surface or subsurface (hypodermic, inter) flow. Typical depletion curves may be abstracted from the analysis of many hydrographs, either in graphical or analytical form. The graphical procedure is advantageously plotted on semilogarithmic paper.

The analytic depletion equation is

$$Q_i = Q_{i-1}K,$$

Q_i = discharge in time i,
Q_{i-1} = discharge in time i minus one time unit, and
K = the depletion factor, inferior to 1,
or in a general form

$$Q_i = Q_{i-t}K^t,$$

where t is the number of time units elapsed between discharges Q_i and Q_{i-1}.

A similar form used in the USSR and also in Czechoslovakia is

$$Q_i = Q_{i-t}e^{-Kt^n},$$

where e = base of natural logarithms,
n = a geographical parameter.

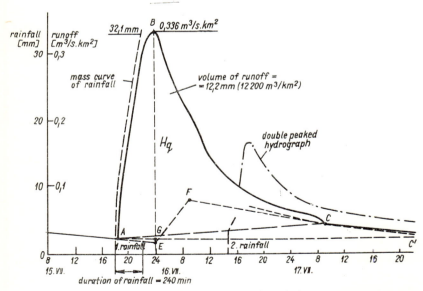

Fig. 4.26. Flood hydrograph analysis

The values of K and n may be ascertained by double logarithming of the equation, and its plotting in a double-logarithmic paper; with values of Q_i and Q_{i-t} from observed hydrographs. Thus, the equation will be $\log (\log Q_i - \log Q_{i-t}) = \log m + n \log t$ where $m = K \log e$. An average value of K and n is computed from straight-line fitting of the depletion curve by inspection in the double logarithmic paper. The procedure is similar to that used in *problem 3* (*Fig. 4.2b*).

Another part of the hydrograph is the so-called rising limb (from A to B in *Fig. 4.26*). The period of time between A and B is often considered as the concentration period during a sufficiently long rainfall, that is, the period necessary for an element of water to flow from the most (hydraulically) distant point in the basin to the outflow cross-section of the basin for which the hydrograph is recorded. This assumption is correct, if the beginning of the rainfall is approximately identical with the beginning of the rising limb and if the rain ends at the moment of the peak discharge at the earliest. If this is not so, the concentration time may be defined as the period from the end of the effective rain (rain of greater intensity than the intensity of infiltration into the soil) to the time when the recession limb of the hydrograph (between B and C on *Fig. 4.26*) joins the typical depletion curve of the basin. Thus, the direct surface runoff from rainfall is enclosed by the line ABC but, simultaneously, there occurs the depletion of subsurface water, called the base flow. Point C and the form of the separation line AC must be determined in order to separate the base flow from direct surface runoff.

Base flow separation

The separation line and point C can be determined, either with the help of a known typical recession curve or by inspection from a large number of hydrographs.

An arbitrary separation of the base flow and direct runoff, and determination of time C', can be made by drawing a horizontal line from point A (in *Fig. 4.26* it is line AC'); by this procedure, the volume of surface runoff is obtained without major error, but the period of recession is extended disproportionately to that of rise.

Another method consists in finding the point C of the hydrograph on the recession line where a flat curve approximating a straight line starts.

The ratio of successive discharges in equal times from this point on remains approximately constant (hence, the straight line in the log co-ordinate).

Some authors assume that the actual contribution of the base flow will be approximately indicated by a broken line AEFC (*Fig. 4.26*), in which part EF coincides in time with the crest period, which forms the third component of the hydrograph.

As long as the procedure used for separation is consistent in the analysis of all hydrographs of the same catchment, the selection of the method is quite arbitrary. Simple methods are, however, recommended since the precision of separation is very dubious by any method.

Selection of flood hydrographs

When selecting the hydrographs suitable for an analysis from long-term records, not all rises in flow should be considered. Since no definition of flood is available for this purpose, some simple rule must be used, but this rule will depend on the climatic conditions of the basin.

Under central European conditions, flood hydrographs result from rain or from snow-melt combined with rain. They are therefore generally distributed throughout the year.

As a rule of thumb for smaller basins in these climatic conditions the base above which all flow rises are considered as *HQ* is the daily maximum flow (in a long-term period) caused by snow melting during a day without rain.

These considerations apply to partial duration series which are recommended for small basins. For annual series, it is irrelevant whether only snowmelt or rain floods or both will be included in the series.

Partial duration versus annual series

When using partial duration series, the recurrence interval (indicated in years) is not connected with the probability by the simple relation $T = \dfrac{1}{P}$, since several occurrences in one year may be selected. The notion of annual occurrence S_r is therefore introduced, similar to the one introduced for frequency analysis of rain intensities (*see page 183*).

$$S_r = \frac{m}{n} = \frac{1}{N}, \qquad (4.66)$$

where m = number of occurrences in n years and the relation between S_r and probability P per cent is

$$P = 1 - e^{-S_r} \times 100, \qquad (4.67)$$

where e = the base for the natural logarithm. Use *Table 36* for a rapid conversion of S_r to P and average recurrence interval of N years.

Table 36

Annual frequency S_r	$P = 1 - e^{-S_r}$ [per cent]	$N = \dfrac{1}{S_r}$ [years]	S_r	$P = 1 - e^{-S_r}$ [per cent]	$N = \dfrac{1}{S_r}$ [years]
0.001	0.1	1000	0.50	39.3	2.00
0.01	0.995	100	0.60	45.1	1.67
0.02	1.98	50	0.70	50.3	1.43
0.03	2.96	33.3	0.80	55.1	1.25
0.04	3.92	25	0.90	59.3	1.11
0.05	4.88	20	1.00	63.2	1.00
0.06	5.82	16.7	1.50	77.7	0.67
0.07	6.76	14.3	2.00	86.5	0.50
0.08	7.69	12.5	3.00	95.0	0.33
0.09	8.61	11.1	4.00	98.2	0.25
0.10	9.52	10.0	5.00	99.3	0.20
0.20	18.1	5.0	6.00	99.8	0.17
0.30	25.9	3.33	10.00	100.0	0.10
0.40	33.0	2.50			

Problem 23

The partial series in *Table 37*, arranged in descending order, includes $133 Hq$ (peak discharges per unit area of basin) observed during 27 years.

Find the annual probability and the recurrence interval for an Hq (peak discharge) of 200 l/s km^2, which was exceeded 48 times during that period and for a peak discharge of 800 l/s km^2, which has been exceeded twice.

Table 37

No.	Year	Hq > 800 l/s km^2	Number of peaks exceeding a flow of [l/s km^2]								Hq annual maximum
			800	700	600	500	400	300	200	100	
1	1928							2	2	4	355
2	1929			1	1	1	1	1	1	3	702
3	1930				1	1	1	1	3	10	680
4	1931					1	1	2	5	9	572
5	1932									1	147
6	1934									1	160
7	1935									9	173
8	1936								1	4	240
9	1937					1	1	1	3	12	570
10	1938								3	7	255
11	1939	1410	1	2	2	2	2	4	5	10	1410
12	1940					1	2	3	4	8	510
13	1941							1	3	7	352
14	1942									2	175
15	1943					1	1	1	2	3	570
16	1944							2	2	7	390
17	1945									2	143
18	1946									2	130
19	1947								1	1	225
20	1948							1	1	2	380
21	1949				2	2	2	2	2	4	670
22	1950								5	9	265
23	1951							1	2	4	390
24	1952									2	125
25	1953						1	1	2	4	470
26	1954									2	160
27	1955	1040	1	1	1	1	1	1	4	4	1040
		$S = 1$	2	4	7	11	13	24	51	133	

Use eq. (4.66)

$$S_{200} = \frac{48}{27} = 1.78,$$

$$S_{800} = \frac{2}{27} = 0.07,$$

$$N_{200} = \frac{1}{1.78} = 0.56, \qquad N_{800} = \frac{1}{0.07} = 14.$$

Use *Table 36* to obtain P:

$$P_{200} \simeq 80 \text{ per cent}, \qquad P_{800} = 6.76 \simeq 7 \text{ per cent}.$$

$Hq = 200$ l/s km² has an average recurrence interval of 0.56 years. Its annual probability is 80 per cent. $Hq = 800$ l/s km² has an average recurrence interval of 14 years and its probability is $P \simeq 7$ per cent.

Frequency analysis of peak flows (annual series)

In extrapolation by frequency analysis of annual series, data from field surveys are often used; these surveys ascertain the marks of past floods and compute the discharge by the slope-area method (*see page 213*).

The arithmetic average MHQ (average peak flood discharge) of series including one such variate HQ_T of a T-year recurrence interval, greater than the period covered by the series, may be computed from the equation

$$MHQ = \frac{1}{T}\left(HQ_T + \frac{T-1}{n}\sum_1^n HQ_i\right)$$

and the coefficient of variation $C_v = \dfrac{\sigma}{MHQ}$

$$C_v = \sqrt{\frac{1}{T-1}\left[\left(\frac{HQ_T}{MHQ} - 1\right)^2 + \frac{T-1}{n}\sum_1^n \left(\frac{HQ_1}{MHQ} - 1\right)^2\right]},$$

where $n =$ number of years in the annual series.

The USSR standard SN2-57 requires that the design peak flood flow include a safety factor

$$\Delta HQ_p = \frac{aE_p}{\sqrt{n}} HQ_p,$$

where E_p = standard deviation of the frequency factor, given in graphs, depending on the type of distribution and the required probability p,

a = factor of gauging of the basin, for a well-gauged basin $a = 0.7$, for a sparsely gauged basin $a = 1.5$ (particularly for unreliable stage-discharge relations).

The above computation of peak flood discharges has been used by the author on the rivers in the Tigris basin in Iraq. The Pearson Type III frequency distribution curve was used for extrapolation of the annual series of HQ. Its comparison with the unit hydrograph method, which is discussed below, and was applied to the same basins in order to obtain 'probable maximum flood' (PMF) with a hydrometeorological input of 'probable maximum precipitation' (PMP) has given the following results:

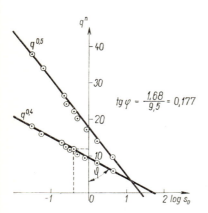

Fig. 4.27. Goodrich distribution (Alexeev)

The 75 per cent of PMF used normally as design flood corresponded, in most cases, closely to peak flows ascertained by frequency analysis extrapolation as having a 1000-year recurrence interval. The 50 per cent of PMF peak corresponded to the 100-years recurrence interval flows.

Frequency analysis of peak flows (partial duration series)

For *partial duration series* and for smaller basins in particular, *the theoretical frequency distribution of Goodrich* is often used. In the version adapted by Alexejev (1955), the equation of this distribution is (*Fig. 4.27*):

$$S_r = S_0 10 \left(\frac{-(HQ - HQ)_0}{\alpha} \right)^n, \qquad (4.68)$$

where S_r = the average number of occurrences in one year,

HQ_0 = value of base flood peak, smaller peaks not being considered, n, S_0, and α being parameters of the curve.

S_0 has a physical significance as the average annual number of all discharges larger than a given HQ. The parameter n may be accepted as 0.4 to 0.5 for smaller basins. The basic equation may be expressed logarithmically:

$$\log S_r = \log S_0 - \left(\frac{(HQ - HQ_0)}{\alpha}\right)^n$$

$$= \log S_0 - (HQ - HQ_0)^n \left(\frac{1}{\alpha}\right)^n. \tag{4.69}$$

The variables are then $\log S_r$ and $(HQ - HQ_0)^n$.

For smaller streams, where HQ_0 may be neglected, the basic relation is then

$$\log S_r = \log S_0 - \left(\frac{1}{\alpha}\right)^n HQ^n. \tag{4.70}$$

To solve this equation, a linear relation is plotted between $\log S_r$ and HQ^n. From it, it is possible to determine $\left(\dfrac{1}{\alpha}\right)$ and subsequently α, S_0, and all the parameters of the equation. To extrapolate for recurrence intervals N beyond the observed period, the equation is then transformed to the form

$$HQ_N = \alpha(\log S_0 + \log N)^{1/n}. \tag{4.71}$$

Problem 24

Calculate peak flows (Hq) of recurrence intervals of 1, 2, 5, 10, 20, 50, and 100 years for a stream with a catchment area of 4.09 km² on which 133 flood peaks Hq were gauged (*see problem 23*) during 27 years. The values of occurrences are given in *Table 37* according to magnitudes of Hq (flood peak discharge per unit of basin area) exceeding 100, 200, 300, 400, 500, 600, 700, and 800 l/s km².

Table 37 indicates that MHq (highest peak flow in the period) was 1410 l/s km². HQ_0 is neglected and values of 0.4 or 0.5 are considered for n.

Further calculations are given in *Table 38*. The relation between log S_0 and Hq^n, according to eq. (4.70), is plotted in *Fig. 4.27* (for $n =$ $= 0.4$ and 0.5). The Figure shows that the linear relation is more suitable for $n = 0.4$. A straight-line relation was fitted by the procedure indicated in problem 3.* For $n = 0.4$ (*see Fig. 4.27*)

$$\alpha = \left(\frac{1}{0.177}\right)^{1/0.4} = \left(\frac{1}{0.177}\right)^{2.5} = 5.55^{2.5} = 76.$$

Table 38

Hq	S	$S_0 = \dfrac{S}{n} = \dfrac{S}{27}$	log S_0	$Hq^{0.5}$	$Hq^{0.4}$
1410	1	0.037	−1.432	37.5	18.1
1040	2	0.074	−1.131	32.2	16.1
700	4	0.15	−0.824	26.4	13.8
600	7	0.26	−0.585	24.5	12.9
500	11	0.41	−0.398	22.3	12.0
400	13	0.48	−0.319	20.0	11.0
300	24	0.89	−0.051	17.3	9.8
200	51	1.89	0.276	14.1	8.3
100	133	4.93	0.693	10.0	6.3
			$\Sigma - \dfrac{3.771}{9}$ $= -0.419$	$\Sigma \dfrac{204.3}{9} = 22.7$	

From the same graph, log $S_0 = 1.68$ and hence eq. (4.70) with substituted parameters α, S_0, and n, is

$$Hq_N = 76.8(1.68 + \log N)^{2.5} \qquad [\text{l/s km}^2].$$

1, 2, 5, 10, 20, 50, and 100 are substituted for N, and corresponding Hq_n are calculated.

HQ_N is computed from Hq_N by multiplying the values of Hq_N by the basin area A:

$$HQ_N = Hq_N A = 4.09 Hq_N.$$

The computation is in *Table 39*.

---

* In the graph in Fig. 4.27 the line fitted for $n = 0.4$ must pass through the point co-ordinates log $S_0 = 0.404$ and $Q^{0.4} = 12.0$ (*see problem 3*).

Table 39

N	1	2	5	10	20	50	100	Recurrence interval in N years
Hq_N	281	427	669	903	1190	1605	1997	(l/s km^2)
$HQ_N = HQ_N A$	1.15	1.75	2.74	3.71	4.87	6.56	8.17	(m^3/s)

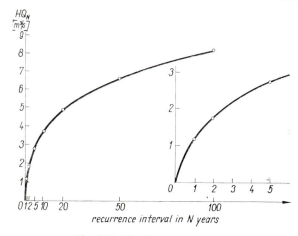

Fig. 4.28. Flood recurrence curve

Recurrence curve of flood peak

In Czechoslovakia and in other countries of central Europe (Germany, Austria) the values of HQ_N are plotted against the recurrence interval N to form the *flood peak recurrence curve (see Fig. 4.28)*. This curve may be plotted directly from a partial duration series, using the annual frequency S_0 and the corresponding recurrence interval N.

Unit hydrograph method

One of the methods for calculating a flood hydrograph with a very short record of data is the 'unit hydrograph' method, very widely used in the United States and in other countries all over the world.

In its basic form, it was first described in 1932 by L. K. Sherman, since when it has been revised, improved, and transposed into nomograms and graphs by many other hydrologists in the United States and in other countries. Thus, there is a wide range of variations on its principle, which may be summed up as follows.

The hydrograph is the sum of the elementary hydrographs from all the sub-areas of the basin, modified by the effect of routing through the stream channels. Since the physical characteristics of the basin-shape, size, slope, channels, etc. are constant, one might expect similarity in the hydrographs which have been caused by rainfalls of similar character.

Two basic assumptions are made.

(a) For a given basin, *the durations* of runoff from rainfalls of the same duration and uniform intensity in time and space *are the same* and do not depend on the total effective rain depth (direct surface runoff volume), it being assumed that the rainfalls are shorter than the time of concentration.

(b) *Volumes* of direct surface runoff within the same time increments *are directly proportional* to the total volumes of runoff.

Thus the 'unit hydrograph' is the hydrograph of a base time duration and of a unit runoff volume. Its volume amounts to 1 in. in the Imperial system of units and, in the metric system, it is 1 cm (10 mm).

The concept of the unit hydrograph is, theoretically not correct. Indeed, no two rainfalls have the same pattern in time and space. In addition, the volume of direct runoff (the volume of the effective rain) has a definite influence on the base time of the hydrograph, particularly in its rising limb. The time of concentration is far from being constant. The author's experiments with physical models of basins and rain simulators (Němec, 1968) demonstrate that, even for two rains of different intensity but in time and space strictly constant, the concentration time is directly proportional to the intensity of rain and to the rate of runoff.

Nevertheless, according to empirical experience, most basins of a size suitable for this method (some authors indicate 5000 km^2 as the upper limit of suitability for the use of the unit-hydrograph method) have a response within the practical margin of accuracy of the computed flood hydrograph.

The unit hydrograph concept is also known under several other names. So, for example, the 'time-area curve' of a basin used by several USSR hydrologists (Kalinin — Miluykov, 1958) is, in substance, a non-dimensional unit hydrograph of the basin.

The practical use of the unit-hydrograph method consists of the possibility of applying a maximum rain depth, either observed in the past or computed (for example, by methods indicated on *page 196* for maximum probable precipitation) to a basin for which the unit hydrograph has either been abstracted from gauged data or synthesized on the basis of catchment characteristics.

Effective net rainfall (rainfall-runoff relation)

In the analysis of the hydrograph on *page 238*, the importance of separating the direct surface runoff from the base flow was emphasized.

In the synthesis of a flood hydrograph a similar but inverse problem must be solved, namely the computation of the volume of direct runoff from the design rainfall (the effective or net rainfall). This may be done by a suitable storm *rainfall-runoff relation.*

The annual precipitation runoff relation (*see page 225*) is not dependent on the soil moisture or infiltration in the soil to such an extent as in the computation of the direct surface runoff from a storm. Thus, to use the unit-hydrograph method, it will be necessary to ascertain the soil moisture before the rainfall, and/or an index (or function) of infiltration during the rainfall.

Antecedent precipitation index (API)

The concept of the 'antecedent precipitation index' (API) has been used with success for this purpose. This method, originated by the US Weather Bureau and combined with a graphical co-axial correlation, has also been applied for purposes of operational hydrological forecasting in Czechoslovakia with considerable success. The method is described in great detail in the book by Linsley, Kohler, and Paulhus, 'Hydrology for Engineers' (McGraw-Hill, 1958), and its principle consists in computing the API according to the equation

$$I_t = I_0 K^t, \tag{4.72}$$

where I_t = API reduced during t preceding days without rain,

I_0 = initial value of API (rainfall depth),

K = a recession factor ranging between 0.85 and 0.98.

This equation indicates that the rate at which moisture is depleted from a catchment is roughly proportional to the amount of storage and the soil moisture decreases logarithmically with time during periods (days) of no precipitation. Since K is a function of potential evapotranspiration, it should be related to seasons or calendar months.

The co-axial correlation, combining correlation of API, duration of storm, and depth of rainfall during the storm with the runoff from it, yields the necessary volume of runoff to be inserted in the unit hydrograph for the computation of the synthetized flood hydrograph.

Infiltration indices and curves

The infiltration approach to the computation of a storm rainfall-runoff relation consists of the application of an infiltration curve or an infiltration index W to the rainfall.

The infiltration curve equation, as originally proposed by Kostyakov (Němec, 1965), has the form

$$i_t = i_1 t^{-a}, \tag{4.73}$$

where i_t = infiltration in time t after beginning of rain [mm/min],

i_1 = infiltration in first unit of time after beginning of infiltration,

a = coefficient, depending on soil characteristics.

i_1 may be ascertained as a function of rain intensity.

Thus, in Czechoslovakia, Dvořák (1962) has proposed a practical equation:

$$i_1 = bi^l, \tag{4.74}$$

where i = intensity of rain,

b, l = coefficients expressing soil characteristics.

It should be noted that Kostyakov's approach is very similar to Horton's infiltration concept and to curves described in the reference above (Linsley et al., 1958).

The infiltration index W represents an average value of infiltration rate throughout the rain,

$$W = \frac{1}{t}(P - R - S),\qquad (4.75)$$

where t = duration of runoff,
$\quad P$ = total volume of rainfall,
$\quad R$ = total volume of runoff,
$\quad S$ = surface retention.

Such an index (or 'runoff coefficient') is mainly derived empirically, on the basis of the regionalization of data on rainfall and resulting runoff; index maps (isolines) have been prepared in Czechoslovakia for purposes of flood flow computation, as indicated later in this book.

In the USSR, extensive field measurements of this index by infiltrometers (*in situ*) have permitted the construction of detailed regionalized maps of isolines of its values.

Synthesis of flood hydrograph (unit-hydrograph method)

Once the value of the effective rainfall is ascertained, the computation of the flood hydrograph by using the unit hydrograph (or distribution graph) is relatively simple.

The procedures are illustrated in *problem 25*.

Problem 25

Determine a unit hydrograph for a gauging site on a stream, for which the flood hydrograph shown in *Fig. 4.26* has been gauged. Use two additional flood hydrographs recorded at the same site, caused by rainfalls of different volumes but of the same duration (240 min).

Using the unit-hydrograph method, calculate the peak flood discharge caused in this basin by rainfall of the same duration, but with a recurrence interval of 100 years, asuming that the average flood 'runoff coefficient' ascertained for the basin is 0.6.

First, the unit hydrograph is derived. The base flow is separated as described above and as indicated in *Fig. 4.26* for all three floods. The time between the beginning of the rain and its peak is about 300 min

for all hydrographs and the total runoff period is around 32 h. (The
hydrographs of the other two floods are already reduced to unit volume
of 1 cm of runoff depth as indicated in *Fig. 4.29*.) The area of the hydrograph in *Fig. 4.26* above the line AC was measured as 85 square units;
one unit is, according to the scale, $0.02\,\text{m/s}\,\text{km}^2 \times 7200\,\text{s}$, or $144\,\text{m}^3/\text{km}^2 =$
0.144 mm (since 1 mm = $1000\,\text{m}^3/\text{km}^2$). The total runoff depth is
hence 85×0.144 mm = 12.2 mm.

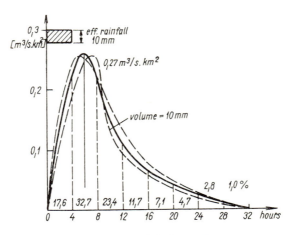

Fig. 4.29. Unit hydrograph

The unit hydrograph will have the same time base, but its area will
represent only 10 mm of runoff. The hydrograph in *Fig. 4.26* may be
reduced by dividing it from the time origin into intervals of four hours.
The ordinates at the end of each interval are then reduced in a ratio

$$\frac{10}{12.2} = \frac{69.5}{85} = 0.83.$$

All three hydrographs have been reduced and the ordinates obtained
are given in *Fig. 4.29*. From these three very similar hydrographs, the
unit hydrograph can be plotted by averaging or selecting the median
value of the ordinates at selected time intervals (the heavy line in
Fig. 4.29). The runoff volume for each 4-h interval from the beginning

of the rainfall, as a percentage of total runoff and in millimetres, is given in *Table 40*.

The maximum ordinate of the unit hydrograph, corresponding to the peak discharge, occurs six hours after the beginning of the runoff; its value is $0.330 \times \dfrac{10}{12.2} = 0.27$ m³/s km². Both the unit hydrograph and the 'distribution graph' are illustrated in *Fig. 4.29*.

Table 40

Hours	0	4	8	12	16	20	24	28	32	Total
Volume [per cent]	17.6	32.7	22.4	11.7	7.1	4.7	2.8	1.0		100
[mm]	1.76	3.27	2.24	1.17	0.71	0.47	0.28	0.10		10

Equation (4.42) is then used to derive rainfall with a recurrence interval of $N = 100$ years and a duration of 240 min. The parameters of the equation for the given basin are: $a = 11.4$, $b = -0.9$, $n = 0.18$.

$$hP \text{ [mm]} = (11.4 \log t - 0.9)\, N^{0.18},$$

$(11.4 \log 240 - 0.9)\, 100^{0.18} = 26.3 \times 2.29 = 60$ mm.

The total rainfall depth is 60 mm. The direct runoff volume is computed with the help of the 'runoff coefficient' indicated above.

$$hSC = 60 \times 0.6 = 36 \text{ mm}.$$

The effective rainfall is thus 36 mm. *Fig. 4.30* shows its distribution in the unit hydrograph. The peak discharge HQ will be $0.27 \times \dfrac{36}{10} = 0.97$ m³/s km².

The derivation of unit hydrographs for any particular duration of rain is possible by the construction of a composite unit hydrograph (the so-called *S curve*).

The S-curve is an integral curve which is obtained by the super-position (addition of ordinates) of several unit hydrographs lagged by a time interval equal to the duration of the unit rainfall *t* with respect to the preceding unit hydrograph. In other words, the *S-curve* is the hydrograph of storm runoff resulting from a continuous uniform effective rainfall intensity lasting indefinitely. With a time base of *T* h for the unit hydrograph, a continuous rain producing 1 cm of runoff in *t* h would produce a constant rate of flow (equilibrium flow) at the end of *T* h. Thus *T/t* h is required to produce an *S-curve* of equilibrium flow. An example of *S-curve* derivation is given in *Table 41a*.

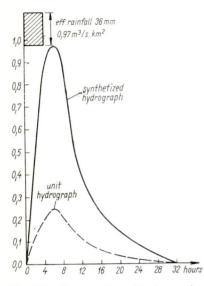

Fig. 4.30. Computation of hydrograph by unit-hydrograph method

If the duration *t* of the unit hydrograph decreases infinitely, approaching zero, the UH becomes the so-called *instantaneous unit hydrograph* (IUH) which is a derivative of the integral *S-curve*. The concept of IUH and its integral *S-curve* advantageously permits representing the IUH and its *S-curve* integral by analytical mathematical equations representing the distribution of runoff as a statistical distribution function. For a detailed description of this approach, the reader is referred to chapter 6, 'Applied Flood Hydrology' by J. E. Nash (Symposium on Flood Hydrology, British Institution of Civil Engineers, 1969).

Synthetic unit hydrographs

Synthetic unit hydrographs, constructed from the characteristics of the catchment, have been proposed by Snyder (1952) and Nash (1958). A theoretical approach to the unit-hydrograph method based on system analysis and resulting in linear models of flood process, was presented by Dooge (1961). Since these techniques of hydrological analysis are beyond the scope of this book, the reader is referred to Chow (1964), Sections 14 and 25.

Table 41a. S-curve example

1	2	3	4	5	6	7	8	9	10	11
time (h)	12-h unit hydrograph	(2) lagged 12 h	(2) lagged 24 h	(2) lagged 36 h	(2) lagged 48 h	(2) lagged 60 h	Total (2) through (7)	(8) lagged 6 h	(8) minus (9)	(10) times 2
6	1						1	0	1	2
12	4						4	1	3	6
18	8	1					9	4	5	10
24	16	4					20	9	11	22
30	19	8	1				28	20	8	16
36	15	16	4				35	28	7	14
42	12	19	8	1			40	35	5	10
48	8	15	16	4			43	40	3	6
54	5	12	19	8	1		45	43	2	4
60	3	8	15	16	4		46	45	1	2
66	2	5	12	19	8	1	47	46	1	2
72	1	3	8	15	15	4	47	47	0	0

The unit hydrograph in particularly suitable for checking the results of flood computations by statistical methods particularly in medium-size basins and in catchments with a short period of flow data. The procedure may easily be computerized.

Formulae for flood peak computations without any direct flow records

In many small basins (for Czechoslovakia, this includes basins from 1 to 50 km^2), it is often necessary to determine HQ without stream-gauging records of any kind.

The use of the so-called 'rational formula',

$$HQ = ikA, \tag{4.76}$$

where i = a given maximum of rainfall intensity,

k = a 'runoff coefficient',

A = basin area.

was once very prevalent in Europe; it is now largely abandoned and reserved only for storm sewer computations.

In small catchments and for the design of small, relatively unimportant projects, the following well-known *exponential formula* is used:

$$Hq_m = \frac{B}{A^n} \Sigma_0$$

or

$$HQ = BA^{1-n} \Sigma_0, \qquad\qquad (4.77)$$

where Hq = the maximum flood peak flow per unit of basin area [m³/s km²],
$\quad HQ$ = the peak flow [m³/s],
$\quad\; A$ = area of basin [km²],
B and n = regional parameters,
$\quad\; \Sigma_0$ = corrections of a local nature for forested areas, slopes, etc.
The author has computed the values of B and n for the main hydrological regions of Bohemia. The advantage of this formula lies particularly in its linearization in a log-log scale paper and in the possibility of comparison of flood peak flows in different climatic and geographic regions by the use of the so-called Creager constant, which is the above parameter B, assuming that $n = 0.5$. The author used this approach advantageously in Iraq, for the catchments of the river Tigris.

A transition from the unit hydrograph method to even simpler methods consists of the substitution of the hydrograph by a triangle, for which the basic time-base components are computed from basin characteristics.

This simple method is sufficiently precise for small catchments and is widely used throughout the world: in the USSR, it has been developed by Sokolovsky; in the USA, it is recommended by the Soil Conservation Service and the Bureau of Reclamation.

The Czechoslovak Hydrometeorological Institute has developed, for use by engineers in the field, a variation of Sokolovsky's formula:

$$HQ_N = \frac{0.28hP_T CAf}{t} \qquad [\text{m}^3/\text{s}], \qquad (4.78)$$

where hP_T = the precipitation depth (for a duration T and of a recurrence interval N) [mm],

C = runoff coefficient during flood period,
t = concentration time [h]:

$$t = \frac{L}{3.6v},$$ (4.79)

where L = the stream length [km],
v = average velocity of flood flow (its approximate values may be obtained from *Table 41b* [m/s],
A = basin area [km²],
f = coefficient of shape of flood hydrograph; its average value is 0.6.
T = duration of the design rainfall is obtained from equation

$$T = (t + 1)^{-0.20} t.$$ (4.80)

Table 41b. Average velocities of flow in the basin for concentration time derivation

Character of basin	Flat	Mildly rolling	Hilly	Highlands	Mountains
	Average slope of basin, from eq. (4.37)				
	0.5 per cent	2 per cent	5 per cent	10 per cent	30 per cent
	velocity [m/s]				
Swampy	0.07	0.08	0.3	—	—
Forested	0.12	0.2	0.5	0.8	1.2
Grassy pastures	0.2	0.5	0.8	1.2	2.0
Gently sloping valley	0.4	0.7	1.0	1.6	2.5
Steep valley	—	—	1.2	2.2	4.0
Rocky steep cliffs	—	—	—	3.0	5.0

This formula, while giving relatively good results, depends very much on the inserted values of the 'runoff coefficient'.

On the other hand, the rainfall recurrence interval gives the possibility of computing a flow of the same recurrence interval, provided that the 'runoff coefficient' assumes maximum values.

In Czechoslovakia, the 100-year flood flow is considered as the basic value. To convert HQ_{100} to HQ_N with $N < 100$, coefficients a_N may be used; they are abstracted from a typical 'flood-recurrence curve' and defined

$$a_N = \frac{HQ_N}{HQ_{100}}.$$ (4.81)

These coefficients were derived for streams in Czechoslovakia by several hydrologists, including the author, and are included in *Table 42*.

Table 42

1	2	3	4	5
N years	Steep non-forested catchment	Partially forested (30—60 per cent) in hills	Forested (60—80 per cent) in hills	Partially forested plains
1	0.06	0.1	0.14	0.18
2	0.08	0.15	0.21	0.29
5	0.13	0.23	0.33	0.44
10	0.21	0.33	0.45	0.55
20	0.34	0.47	0.60	0.67
50	0.62	0.70	0.81	0.84
100	1	1	1	1

If the HQ_{100} is computed, other HQ_N may be derived, using coefficient a_N from *Table 42*, by the equation

$$HQ_N = HQ_{100} a_N.$$ (4.82)

The 'flood-recurrence curve' may be plotted from these values. When computing the peak flow by the above formula, if there are reservoirs in the basin, their retention influence must be taken into account, for example, by simplified flood-routing methods (*problem 30*).

Whenever possible, the computed values should be checked against the largest discharges in the past and ascertained by the slope-area method from flood marks (*problem 17*). The use of the Sokolovsky formula is illustrated below.

Problem 26

Calculate HQ_N in a basin of area $A = 20$ km^2 for recurrence intervals $N = 1, 2, 5, 10, 50$, and 100 years.

The computed discharges will be used for spillway design of a small reservoir. The main stream is 6.2 km long. The mean slope of the stream is 1.3 per cent and the area of the basin is 25 per cent forest. The flood 'runoff coefficient' is 0.4 according to comparison with a similar, gauged stream in the region. Two methods will be used for the computation.

The regional approach by formula (4.77). The stream is in a region for which the following parameters of eq. (4.77) were given:

$$B = 2.9 \quad \text{and} \quad n = 0.3,$$

thus

$$HQ = 2.9A^{1-0.3} \quad [\text{m}^3/\text{s}].$$

$$\log VQ_{100} = \log 2.9 + 0.7 \log 20 = 1.372,$$

$$HQ_{100} = 23.5 \text{ m}^3/\text{s}.$$

Since the forestation of the basin is 25 per cent below the average, HQ_{100} will be raised by 10 per cent:

$$HQ_{100} = 23.5 + 2.5 = 26 \text{ m}^3/\text{s}.$$

Computation according to formula (4.78). The values of P_T, C, f, t, and auxiliary values v and T must be computed first.

The value of v is taken from *Table 41b*. For a 25 per cent forested basin with 1.3 per cent slope of the main stream $v = 0.6$ m/s. From eq. (4.79):

$$t = \frac{L}{3.6v} = \frac{6.2}{3.6 \times 0.6} = 2.9 \text{ h}.$$

From eq. (4.80):

$$T = (t + 1)^{-0.20} t = (2.9 + 1)^{-0.20} \times 2.9 = 2.2 \text{ h} = 132 \text{ min}.$$

The value of P_T for $N = 100$ is obtained from eq. (4.42) with the following parameters $a = 12.1$, $b = -0.78$, and $n = 0.20$ (*Table 19*):

$$P_T \text{ [mm]} = (12.1 \log 132 - 0.78) \, 100^{0.2} = (12.1 \times 2.12 - 0.78) \, 2.52 =$$
$$= 63 \text{ mm}.$$

The average value of f is 0.6; C is 0.4. Substituted in eq. (4.78):

$$HQ_{100} = \frac{0.28 \times 63 \times 0.4 \times 20 \times 0.6}{2.9} = 29.2 \text{ m}^3/\text{s}.$$

Thus, two different values have been obtained for HQ_{100}: 26 m³/s and 29.2 m³/s. Since a spillway design is involved, the safer value of 29 m³/s

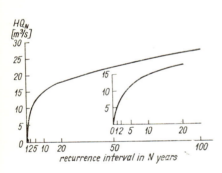

should be retained. To derive HQ_N for $N = 1, 2, 5, 10, 20,$ and 50, eq. (4.82) is used in which coefficients a_N are chosen from *Table 42*, depending on the character of the basin. The 'flood-recurrence curve' is then computed and plotted in *Fig. 4.31*. For HQ_{50}, for instance, $a_N = 0.84$ and thus, according to eq. (4.82), $HQ_{50} = 26.5 \times 0.84 = 22$ m³/s. The computation is in *Table 43*.

Fig. 4.31. Flood recurrence curve

It should be stressed that the choice of coefficients in eq. (4.78) is often difficult and subjective. From this standpoint, the results may only be considered as approximate. Greater mathematical accuracy in calculations is thus unnecessary and the values may be rounded off to a whole m³/s.

Table 43

N	1	2	5	10	20	50	100
a_N	0.18	0.29	0.44	0.55	0.67	0.84	1
HQ_N [m³/s]	5.2	8.5	12.8	16.0	19.6	24.5	29.2

4.9   Minimum flow

On *page 38* (section 2.4c), the different characteristics of minimum flow are given with their definitions. As long as there is a long-term time series of stream-gauging data, the computation of these characteristics, and in particular of NNQ and $MNQ(y)$, is self-evident from their definition. The daily flow duration curves are most useful for this purpose.

If the flow duration curve of mean daily flows is available for an average year, MQ_{355} and MQ_{364} can be read directly from it and thus $MLQ(y)$ (average lowest annual flow) ascertained. If these curves have been plotted for a series of years, it is possible to abstract a series of Q_{355} and Q_{364} and process it by frequency analysis; for instance, by plotting a frequency distribution of minimum flows. The procedure will be similar to that in *problem 7, 8,* or *25.* However, MQ_{355} and MQ_{364} and MQ are without specified frequency of occurrence. If, for economic studies, this must be obtained, at least approximately, and there are no records for several years — as, for instance, in small basins — it is possible to obtain Q_{355} and Q_{364} of a certain probability p per cent by using the ratio

$$\frac{MQ_{355}}{MQ(y)} = k_{355} \quad \text{and} \quad \frac{MQ_{364}}{MQ(y)} = k_{364}.$$

If MQ_{355} and MQ_{364} have not been given they may be taken from regionalized curves, using eq. (4.64). $MQ(y)_{p\%}$ may be computed for $p = 80, 85, 90,$ and, in exceptional cases, 95 per cent. The approximate value of MQ_{355} and MQ_{364} for a certain p will then be $MQ(y)_{p\%} k_{355}$ and $MQ(y)_{p\%} k_{364}$. It should be noted, however, that the ordinates of the flow duration curves may not all be reduced by the same ratio for a given p. The largest reduction is in the interval from Q_{60} to Q_{210}. In *Table 44*, this ratio is given for reduction of the flow duration curve Q_n from $MQ(y)$ of a probability $p = 80$ per cent (for a small stream).

Problem 27

Compute minimum flow Q_{355} of a probability $p = 80$ per cent for the stream in *problem 20.* $MQ(y) = 19.3 \text{ l/s}$; $Q(y)_{80\%} = 6.8 \text{ l/s}$. In

Table 44

n [days]	5	10	30	60	90	120	150	180
$\dfrac{Q_{n\,90\%}}{MQ_n}$	0.59	0.51	0.47	0.37	0.34	0.29	0.28	0.28
n [days]	210	240	270	300	330	355	364	
$\dfrac{Q_{n\,90\%}}{MQ_n}$	0.30	0.37	0.39	0.40	0.43	0.50	0.50	

problem 22, the flow duration curve of $MQ(d)$ was obtained for $MQ(y)$ in which $Q_{355} = 2.5$ l/s and $Q_{364} = 1.32$ l/s.

$$k_{355} = \frac{2.5}{19.3} = 0.13, \qquad k_{364} = \frac{1.32}{19.3} = 0.07,$$

$$Q_{355} \text{ for } p = 80 \text{ per cent} = 6.8 \times 0.13 = 0.88 \text{ l/s},$$

$$Q_{364} \text{ for } p = 80 \text{ per cent} = 6.8 \times 0.07 = 0.48 \text{ l/s}.$$

For the real-time forecasting of low flows, recession curves, as described in connection with hydrograph separation (*see page 237*), are being used. These curves assume, however, the availability of direct gauged data on flow.

If only an approximate value of $MQ(y)$ is available, empirical equations are used to determine MLQ. M. E. Shevelev's equation may be used (A. A. Lutcheva, Prakticheskaya gidrologia, Leningrad, 1950; page 236):

$$MLQ(\text{m}) \, kA^{0.034} MQ(y)^{0.94}, \tag{4.83}$$

where $MLQ(\text{m})$ = the average minimum monthly discharge [l/s],

$MQ(y)$ = average annual runoff [l/s],

A = catchment area [km²],

k = regional coefficient (its average value is 0.155).

Problem 28

Compute the average minimum monthly discharge $MLQ(m)$ for the basin in *problem 20*.

$$MQ(y) = 19.3 \text{ l/s},$$

$$A = 13.9 \text{ km}^2.$$

From eq. (4.83):

$$MNQ(m) = 0.155 \times 13.9^{0.034} \times 19.3^{0.94},$$

$\log MLQ(m) = \log 0.155 + 0.034 \log 13.9 + 0.94 \log 19.3,$
$\log MLQ(m) = -0.810 + 0.034 \times 1.143 + 0.94 \times 1.286 = 0.4374,$
$\quad MLQ(m) = 2.74 \text{ l/s}.$

The calculated value corresponds to Q_{355} from the flow duration curve of $Q(d)$ for an average year.

5 Elements of Flow Control and Routing

5.1 Problems of flow control

The term '*control*' suggests that a positive influence of man changes the natural random element of surface runoff in a catchment and the resulting flow in the channel of a stream in a favourable way for man's interest.

The techniques of changing or conserving the natural condition in a catchment in favour of human interests are examined in the field often known as 'watershed management' and can be either of a biological or of an engineering nature. They will not be described here but some indications of the principles on which they are based are given in section 2. These techniques almost always require long-term planning and the results, although of a durable character, are often slow to materialize.

Engineering interventions in the natural regime of flow in the channel are more effective and rapid, although not always durable and sometimes accompanied by unforeseen negative side effects. Such interventions most often consists of reservoir building by means of dams, and this on large as well as small streams. The purpose of reservoirs may be multiple; in fact, multi-purpose projects are most advantageous from an economic point of view.

The activity involved in this type of flow control is often called 'water resources design (management)' and contains elements of hydraulic and other engineering as well as economics. Hydrology is always the first and basic component of it. The inter-relationship of hydrology with engineering and economics in this respect is so close that a basic hydrology book such as this one can only contain very elementary approaches to problems of the hydrology of flow control.

While the storage in man-made reservoirs represents the most important element involved, it is evident that storage also occurs in natural channels and flood plains, particularly during flood periods. Therefore, the transformation of flow by the channels and other natural storage, known under the name of 'flow or flood routing', is often included under the subject of flow control, and flood routing through man-made reservoirs is only a particular and often the simplest case of the general routing problems.

The complexity of the natural conditions of the catchment, of the natural channel storage influence on the flow, and its combination with the effects of man's activity and the considerable economic impact of flow control have given an impetus to an accelerated development of highly sophisticated modern systems for collection, transmission, and processing of hydrological data.

This last chapter is, therefore, devoted to considering some elementary aspects of the hydrology of:

(a) flow control by reservoirs;
(b) flow (flood) routing through natural or man-made storage;
(c) development of modern data systems in hydrology.

5.2   Elements of reservoir design

The flow control by a storage reservoir may assume, basically, two functions: increase of flow in periods of low flow and decrease of flow in time of floods.

The inflow into the storage is most often not controlled, and so the input is a function of the natural regime runoff conditioned by the character of the catchment and its climate.

If time redistribution of flow is considered to be the main purpose of flow control, the following types of control may be envisaged or needed:

(a) long-term year-to-year carry-over control;
(b) annual and seasonal control;

(c) weekly control;

(d) daily control;

(e) irregular—for example, the so-called 'buffer' control where a storage down-stream absorbs and redistributes large flows released irregularly by an upstream storage, after their use for hydropower production.

The graphs of redistribution of flows (control) mentioned above are in *Fig. 5.1a.*

From the graphs of inflow, it is clear that the flow control design is based on the natural hydrological regime—depending on the nature of control, the hydrograph of inflow is a long-term multi-annual hydrograph of annual average flows $Q(y)$, or monthly average flows $Q(m)$, or finally of daily flows $Q(d)$. For evaluation of the routing effect of the storage for floods of shorter duration, a continuous hydrograph is most advantageous although short-interval hydrographs (hourly, three-hourly) may also be used.

The necessary hydrological data for the hydrograph derivation may be taken directly from the past—a long-term representative sequence of data recorded according to the calendar may serve as the design period.

Methods using such an approach are relatively simple and have been used for many years. However, they lack the possibility of an objective economic evaluation of the effect of control—the probability of the failure of the storage to meet the demand on increased or decreased flow. To qualify the risk intrinsically present in the extrapolation of past flow data into the future, probabilistic methods must be used, basically similar to those indicated in section 4.3 but with new elements added. If the calendar sequence of flow is to be predicted in the future with a certain probability of occurrence, consecutive variates may no longer be considered as completely independent. In addition to their random element, they will contain a deterministic component. This deterministic influence of the preceding variate is more attenuated, the more its occurrence is distant in time. Thus, artificial probabilistic sequences of flow are simulated on the basis of parameters derived from past observed data; such simulation is the substance of *stochastic* processes used in flow control design. The simulation of series of time-dependent flows uses mainly the so-called Monte Carlo method, which consists of creating

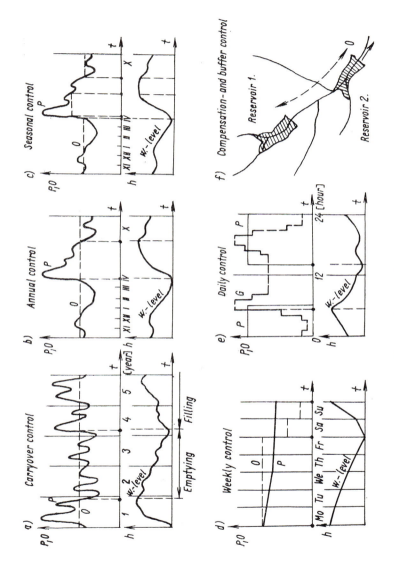

Fig. 5.1. Types of flow control

either one very long artificial series of annual flows (up to 5000 years) or a large number (100 or more) of shorter time sequences (of 50 years each). It is not within the scope of this book to give the details of derivation of such series. The reader is directed to the following references: V. T. Chow, Section 8−4 and 25 (1964), Klemes (1967, 1968), Gould (1961), Moran (1959), Lloyd (1969), Kritskii and Menkel (1952), Svanidze (1964), Savarenskij (1951).

Whether a classical method using past records or a simulated stochastic process is used, the function of the storage may be analysed by relatively simple water-balancing methods. These are either digital or graphical. These last are most often used for simple engineering purposes, particularly in connection with small reservoirs or dams. The basic design of the reservoir requires the knowledge of the volume of storage needed for a certain control of flow. Such a problem may be solved by a numerical balance of the inflow and outflow or by a graphical representation in the mass curve of flow.

The mass curve is based on a calendar sequence of average daily, monthly, or annual flow or their graphical representation in a bar graph.

This graphical derivation is based on the equation:

$$W = Qt \quad \text{and} \quad \frac{\mathrm{d}W}{\mathrm{d}t} = Q \quad \text{or} \quad \frac{W_2 - W_1}{t_2 - t_1} = Q(t_2 - t_1),$$

where W = volume of flow [m³],
Q = flow (discharge) [m³/s],
t = time [s].

The procedure is illustrated in *problem 29*.

Problem 29

A small reservoir is to be built on a stream. The average annual and monthly discharges of the stream were ascertained in *problem 21*. The reservoir is supposed to have a volume sufficient to control a constant monthly average outflow of 15 l/s throughout the year.

The mass curve of average monthly flows is ascertained first by computation. The average monthly flows $MQ(\mathrm{m})$ [l/s] are given in column 2 of *Table 45*. By multiplying these by the number of seconds in each month (*see problem 21*), the average monthly volumes of flow are computed (column 3 of *Table 45*). The increments of volumes of flow are added in column 4 of *Table 45*.

Table 45

1	2	3	4	5	6		7	8
Month	Monthly flow rates [l/s]	Monthly volumes [m³ × 10³]	W [m³ × 10³]	Average annual flow summation	Residual curve [m³ × 10³]		Residual curve for annual average summation 4−5	Notes
					+	−		
XI	17.3	45.7	45.7	50.8	—	5.1	−5.1	
XII	17.3	45.7	91.4	101.5	—	5.1	−10.2	
I	19.7	51.8	143.2	152.3	1.0	—	−9.2	
II	29.0	76.1	219.3	203.0	25.3	—	+16.1	
III	34.7	91.4	310.7	253.8	40.7	—	+56.8	
IV	32.4	85.3	396.0	304.5	34.6	—	91.4	
V	17.3	45.7	441.7	355.3		5.1	86.3	
VI	11.6	30.4	472.1	406.0		20.3	66.0	
VII	15.1	39.6	511.7	456.8		11.2	54.8	
VIII	11.6	30.4	542.2	507.5		20.3	34.5	
IX	10.4	27.4	569.6	558.3		23.3	11.2	
X	15.1	39.6	609.1	609.1		11.2	0.0	
	231.5	609.1	$\dfrac{MQ(y)}{12} = \dfrac{609.1}{12} = 50.8$		+101.6	−101.6		

Column 8 notes: $MQ(y) = \dfrac{231.5}{12} = 19.3 \text{ l/s}$

This column, plotted in rectangular co-ordinates, represents the mass curve of average monthly flows (*Fig. 5.2*). The final ordinate of the curve (*AB*) is 609.1 × 10³ m³; it is the total annual volume of flow. The average yearly flow $MQ(y)$ is expressed in column 5 of *Table 45*. Its graphical representation (*Fig. 5.2*) is a straight line joining the origin of co-ordinates to the final point of the mass curve of flow A; the tangent of angle α (*see Fig. 5.2*) is

$$\tan \alpha = \frac{AB}{OB} = \frac{W \text{ (year)}}{t \text{ (year)}} = \frac{609.1 \times 10^3}{365 \times 86\,400} = 0.0193 \text{ m}^3/\text{s} =$$

$$= 19.3 \text{ l/s} = MQ(y).$$

For graphical solution, the hydrograph is plotted as a bar graph.

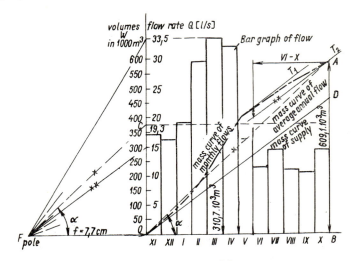

Fig. 5.2. Mass curve of flow

The graphical solution is based on eq. (5.1):

$$\tan \alpha = \frac{Q}{1} = \frac{W}{t}, \qquad \text{hence} \qquad W = t \tan \alpha. \qquad (5.1)$$

If an arbitrary pole is chosen on the axis of abscissae (time) (F in Fig. 5.2), the rays leading from this pole and passing through flow values Q on the ordinate (discharge) axis will form angles whose tangent will correspond to discharges Q because each tangent $= \dfrac{Q}{f}$ where f is the constant polar distance. Since this polar distance is arbitrary (its value influences only the scale in which W will be obtained), it may be assumed that it is equal to a unit of the scale and hence

$$\tan \alpha = \frac{Q}{1} = Q.$$

The construction for the flow for March is shown in *Fig. 5.2* (i.e., the highest average monthly flow). Its value 33.5 l/s is marked on the ordinates (discharge) axis and connected to the pole *F*. A parallel line to this ray is drawn from the point where the February flow ends. At the point of its intersection with the ordinate indicating the end of March is the final point of the mass curve representing the total volume of flow from 1 November to 31 March, which is 387.7×10^3 m^3.

It is, however, necessary to ascertain the scale for the volumes of flow *W*. If the pole distance is chosen arbitrarily, this scale is computed from the equation

$$f = \frac{m_w}{m_t m_q}, \qquad (5.2)$$

where f = the pole distance [cm],

$\quad m_q$ = scale of flows (number of discharge units [m^3/s] in 1 cm),

$\quad m_w$ = scale of volumes (number of units of volumes [m^3] in 1 cm),

$\quad m_t$ = time scale (number of time units [s] in 1 cm).

Then according to eq. (5.2),

$$f = \frac{50 \times 10^3}{86\,400 \times 30 \times 0.0025} = 7.7 \text{ cm.}$$

All scales must be homogeneous in units (in other words, if m_q is in m^3/s, m_w must be in m^3 and m_t in s).

The polar distance (and the pole) can be advantageously constructed graphically. if the final co-ordinates of the mass curve and the scales are known.

The construction of the mass curve is self-evident from *Fig. 5.2*. If it is necessary to construct a mass curve for a longer period with sufficient accuracy in all its scales, this cannot be accomplished in rectangular co-ordinates on graph paper of the usual size.

The mass curve is, therefore, often drawn in oblique co-ordinates. These co-ordinates are selected so that the final point of the mass curve is on the same horizontal line as its origin.

The arithmetical solution is to subtract the mass curve of the average flow from the mass curve of incremental flows.

This is done in columns 6 and 7 of *Table 45*.

In the reduced mass curve plotted from column 7 of *Table 45*, the flow volume values must be read at a new axis turned by an angle β from the original position. Because the abscissae (time) axis does not turn but remains in its original position, an oblique system of co-ordinates is thus created in which the co-ordinates have an angle of $(90° - \beta)$ or $(90° + \beta)$. The value of β can be computed from the equation

$$\tan \beta = MQ \frac{m_t}{m_w}, \qquad (5.3)$$

where β = the angle of the turned axis of volumes,
  MQ = average discharge during the given period,
  m_w = scale of volumes (number of volume units in 1 cm),
  m_t = time scale (number of time units in 1 cm).

The new scale of volumes (in oblique co-ordinates) will be

$$m_w \text{ (oblique)} = m_w \cos \beta.$$

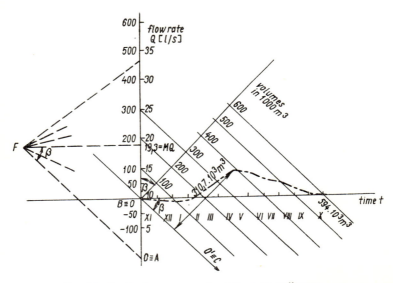

Fig. 5.3. Residual mass curve in oblique co-ordinates

The graphical solution is much simpler, however, and is used almost exclusively. The pole is elevated vertically so that it lies on the same horizontal level as the value of the average discharge plotted on the discharge scale (*see Fig. 5.3*). Angle β will then be the angle between this horizontal line and the ray joining the pole with origin (zero) of the discharge scale. In this way, the direction of the axis of volumes is also obtained. The scale of volumes is obtained by projection of the scale in the rectangular system in the direction of the new axis. For easier reading of the volume values, parallel lines with this new axis are drawn in rounded-off values of the scale; in *Fig. 5.3*, it is for values of 100, 200, and 300 × 10^3 m^3.

The plotting of a mass curve in oblique co-ordinates is identical to that in rectangular co-ordinates. It should be kept in mind, however, that the abscissae (time) axis remains horizontal and hence the time limits are perpendicular to it, i.e., vertical. The construction is shown in *Fig. 5.3*.

If the mass curve has been plotted, the remainder of problem 30 can be solved. The volume of the storage in the reservoir which would control an average outflow of 15 l/s is to be ascertained. Projecting in *Fig. 5.2* the outflow of 15 l/s on the flow scale from point O, the mass outflow curve is plotted (line OD in *Fig. 5.2*), parallel with the ray of 15 l/s, indicated with two asterisks.

Parallel with this line, two tangent lines are drawn above and below the mass curve (lines T_1 and T_2 in *Fig. 5.2*). The upper tangent touches the curve at a point which corresponds to the beginning of the month of June and the lower tangent at a point corresponding to the end of October. This indicates that, in this period, the average monthly inflows are lower than 15 l/s (see also column 2 in *Table 45*). The volume of water which is to be stored in the reservoir in order to supplement the outflow of 15 l/s, is given by the vertical distance between tangents T_1 and T_2. This distance is measured on the volume scale; the volume of the necessary storage in the reservoir is 35 000 m^3.

It should be pointed out that the storage was determined by using average monthly flows; the variation in flow during the months was not taken into account. A more accurate solution would require the use of average daily flows. The procedure, however, would be exactly the same, in the oblique co-ordinates as well.

5.3 Flood routing

The movement of water in the channel has been described in section 2.3. It is evident that, during the movement of a flood wave through a channel reach, the most complex type of flow occurs – the non-uniform unsteady flow.

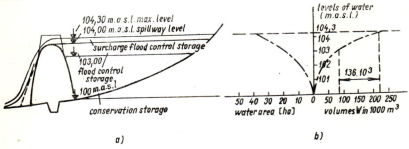

Fig. 5.4. Level-area and level-volume curves of a reservoir

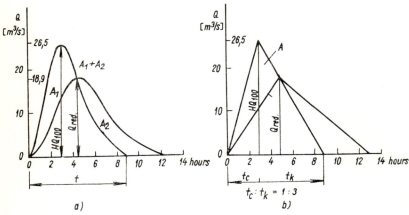

Fig. 5.5. Flood routing by reservoir

Although, as mentioned above, the hydraulic substance of the pheno- menon is basically the same, both for a man-made or natural reservoir or a reach of the river channel, with respect to the flood hydrograph two quite different effects may result. The storage of water in the reservoir is at first controlled (*see Fig. 5.4a*); once the reservoir is full, the *pondage* storage occurs only in the pool above the controlled volume of the reservoir (above the spillway level) and the effect of detention (lag) of flow transforms the inflow into an outflow hydrograph with a lower peak and a larger time base. As the inflow has to exceed the outflow for any storage to occur, the peak of the outflow hydrograph will occur after the inflow has reached its peak, at the time when outflow equals inflow. Thus, a reservoir, or any other type of pondage storage (resulting from lag of flow in the storage), always attenuates the peak discharge (*see Fig. 5.5a*).

In a reach of a perfectly uniform prismatic channel, the flood wave is only lagged without changing its shape; thus, its peak is not reduced. A real reach of a river, however, combines the effects of a reservoir and of a uniform prismatic channel; in many ways, it behaves as a reservoir and therefore the routing may be computed for it as such.

The routing problem presents three aspects.

(a) With a known outflow hydrograph and known storage to compute the inflow hydrograph. This is particularly so with smaller reservoirs, where a stream-gauging station is below the dam and the hydrograph of natural runoff from the catchment is needed (e.g., for unit hydrograph method).

(b) With a known outflow and inflow hydrograph to compute the storage. This is the case when the storage in a reach of the river channel is ascertained by a comparison of actual hydrographs of inflow into and outflow from the reach. Gauging stations have to be situated at both ends of it.

(c) With a known inflow hydrograph and a known (or approximately assumed) storage, the outflow hydrograph is computed. This is the most frequent aspect of the routing problem. It comprises the design of reservoirs with respect to their effects on the attenuation of the flood peaks, the computation of flood wave transformation along the river in real-time flood forecasting, etc. In simplified terms the same approach may be used for construction of models of surface runoff and thus for the synthesis of hydrographs from a catchment simulated by a series of reservoirs or reservoirs and channels (Nash, 1968).

Since the basic phenomenon in the routing problem is unsteady flow, the differential equations describing this flow may be used for the solution of the problem. These are

$$\frac{\partial H}{\partial x} + \frac{V}{g}\frac{\partial V}{\partial x} + \frac{1}{g}\frac{\partial V}{\partial t} + \frac{V^2}{C^2 R} = 0,$$

$$A\frac{\partial V}{\partial x} + V\frac{\partial A}{\partial x} + B\frac{\partial H}{\partial t} = 0,$$

where H = water depth,
      V = average velocity,
      A = cross-sectional area of channel,
      B = surface width of flow,
      R = hydraulic radius,
      C = Chézy coefficient,
      g = gravity acceleration,
      x = distance along the flow direction,
      t = time.

The first equation is often called the momentum (motion) equation and expresses the conservation of energy in a physical system; the second is called the equation of continuity and expresses the conservation of mass. The exact analytical solution of the equations does not exist; however, they may be solved by assuming some simplifications. As long as finite time and space differences are selected and boundary conditions assumed, the equations may sometimes be solved numerically. Even for relatively simple natural conditions, however, the computations are so laborious that they are only possible with the use of an electronic computer. The method of characteristics (Abbott, 1966) may be used, for example. In less simple cases, larger simplifications are necessary. Instead of the hydraulic approach, a hydrological solution is sought which is based on the above equation of continuity in a simplified finite form, the 'storage equation':

$$I_t = O_t + \frac{\Delta S}{\Delta t},$$

where I_t = mean inflow in period Δt,
     O_t = mean outflow in period Δt,
     S = change of storage in period Δt.

For practical solution (to ascertain the outflow hydrograph), this equation is transformed to

$$\frac{\Delta t(O_1 + O_2)}{2} = \frac{\Delta t(I_1 + I_2)}{2} - (S_2 - S_1),$$

where all are as above, except that subscripts 1 and 2 indicate the beginning and end of a finite routing interval Δt. Many routing methods for the solution of this equation are available. Best-known is the Muskingum method, originated by McCarthy, using the concept of combination of prismatic and wedge storage. The method is described by Carter and Godfrey (1960) and by Wilson (1969). In Czechoslovakia, the methods of Klemes (1960) and Urban (1956), and in the USSR those of Potapov (1952), Kalinin — Miljukov (1958), and others are used. While routine computer programs (software) are used for solution by digital computers, graphical methods are most advantageous for problems of smaller size and importance. Klemes's method for a reservoir (prismatic storage only) is given below.

For purposes of real-time hydrological forecasting where the routing problem is to be solved quickly and in many reaches of a river, the electrical analogy may also be used. Electronic analogue routing computers are based on the assumption that the storage is a linear (or other) function of the inflow or $S = kI$. Then

$$I_t = O_t + k\frac{\mathrm{d}I_t}{\mathrm{d}t},$$

where k = lag time (storage coefficient).

This is an equation of intensity of electric current flow through a circuit with a parallel capacitance C and resistance R (an RC element)

$$I_{et} = O_{et} + RC\frac{\mathrm{d}I_{et}}{\mathrm{d}t},$$

where I_{et} = intensity of flow before the RC unit,
O_{et} = intensity of flow after the RC unit.

An electronic analogue computer then integrates the above equation instantaneously; the values of k (RC) are either assumed and compared to the measured value, or derived directly from inflow and outflow hydrographs. The Department of Water Resources of the Prague Agricultural University, with the author's collaboration, is at present engaged in the comparison and derivation of the storage coefficient k, using both the digital solution of the differential equation of unsteady flow and the

electronic analogy solution of the simplified storage equation (Němec –
Zezulak, 1970).

A simple graphical method for routing through a reservoir has been
worked out by Klemes (1960). The inflow hydrograph $I = f(t)$ is a non-
analytical curve. The outflow O is, however, also a function of the elevation h (stage) of the water in the reservoir, since the reservoir's outlets
and the spillway rating curves are only functions of h. The volume of
water in the reservoir V (or the storage) is also a function of h and may
be ascertained from the level-area and level-volume curves of the reservoir
(*see Fig. 5.4*). The method consists in finding for the outflow $O_i = f(V)$
(at any finite interval i) the corresponding value of $O_i = f(t)$, h and V
in this interval being known from the inflow $I_i = f(t)$. The graphical
solution is illustrated in *Fig. 5.6*. $I = f(t)$ is the inflow hydrograph.
I_d is the value of inflow at the time when the reservoir is full and only
the pondage (lag) storage is considered. $O = f(V)$ may be represented as
the so-called transformation curve. $Q = f(t)$ is the outflow hydrograph.
In *Fig. 5.6*, all three curves are evident. The scales of discharges for O
and I must be the same in both graphs; the scale of V is arbitrary. For
any value $O_x = f(V)$, a corresponding value $\Delta t V_x$ may be computed.
When these values are plotted in the graph of the transformation curve
($O = f(V)$) on the scale of V, the straight (reduction) lines connecting
them to values O_x correspond to the relation $\tan z = \dfrac{V_x}{O_x} = \Delta t$. If only
equal Δt are used, one single 'reduction line' is necessary; it is, however,
recommended that shorter Δt be used in the vicinity of the peak, so that
several 'reduction lines' may be constructed.

The routing then proceeds as follows. The inflow hydrograph is divided
into intervals Δt, the medians of these intervals indicating the values
$I_0, I_2, I_3, \ldots, I_i, \ldots$ The routing starts at $I_0 = O_0 = O_d$. The outflow
hydrograph is constructed graphically from values $O_0, O_1, O_2, O_3, \ldots,$
$\ldots, O_i$. It may start, however, from any value O_i which is known. Thus,
the value of V_i is also known, both being at the end of the interval Δt_i.
A perpendicular erected to the point V_i is intersected by a line parallel to
the horizontal and indicates the value of I_i at point A. From this point,
a line parallel to the 'reduction line' (corresponding to the selected Δt)
is drawn which intersects the transformation curve $O = f(V)$ at point B.

Its ordinate indicates the volume in the reservoir at the end of the interval Δt_{i+1} and its abscissa is the outflow O_{i+1} which is transferred by a parallel to the horizontal to intersect the perpendicular erected in the middle of this interval. Thus, the next point O_{i+1} of the outflow hydrograph has been constructed. The construction is correct $if\,(I_{i+1} - O_{i+1}) \cdot \Delta t_{i+1} = \Delta V_{i+1}$. This is the case in the triangle ABC (*Fig. 5.6*) since $BC = AC \tan \alpha$, or $BC = \Delta V_{i+1}$, $AC = I_{i+1} - O_{i-1}$. $\tan \alpha = \Delta t = \Delta_{i+1}$. The maximum storage V_{max} corresponds to the peak of the outflow hydrograph and represents the area between the curves $I = f(t)$ and $O = f(t)$ in *Fig. 5.6*.

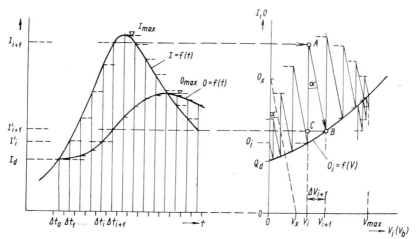

Fig. 5.6. Graphical routing method according to Klemeš

This method is advantageous for routing through a reservoir, for which the transformation (storage) curve may be constructed from the outlet and spillway discharges. For a river reach, such a transformation curve is most difficult to ascertain; the Muskingum or Kalinin-Miljukov procedure of routing, with the storage (lag) k coefficient is then recommended.

On small streams, recorded hydrographs seldom exist, and the solution may then be simplified by replacing both hydrographs with triangles (*Fig. 5.56*). On the basis of this simplification, equations have been derived for an approximate reduction of the peak flow by reservoirs. One such equation is that of D. I. Kocherin:

$$O_{max} = I_{max}\left(1 - \frac{V}{A}\right) \tag{5.4}$$

or

$$O_{max} = I_{max} - \frac{2V}{t}, \tag{5.5}$$

where O_{max} = the peak flood flow of the outflow hydrograph,
   I_{max} = the peak flood flow of the inflow hydrograph,
      V = volume of the flood retention storage (pondage only),
      A = volume of flood (the same before entering the reservoir and after leaving it),
      t = approximate flood duration, or the time base of the flood hydrograph.

The equation assumes that no water is retained in the reservoir for permanent storage. Thus, the areas of both triangles (hydrographs) in *Fig. 5.5b*, representing the volume of the flood, are the same.

Problem 30

In *problem 26*, the peak flood flow *HQ* of 29.3 m³/s was obtained for a stream with a 20 km² catchment. There is a small reservoir on this stream whose level-area and level-volume curves are shown in *Fig. 5.4b*. Find the *HQ* reduced by the routing of the flood through this reservoir.

Equation (5.4) is used for the solution. The flood duration *t* can be determined either on the basis of actual observation of a large flood in the past, or, approximately, by computing the recession time t_k of the flood as a function of the concentration time t_c (*Fig. 5.5b*). This last was obtained in problem 27 as 2.9h.

The ratio t_c/t_k is very unstable. However, the smaller the basin, the greater this ratio usually is: on basins with an area of 50 km², a ratio 1/2 may be assumed, on catchments with an area of 5 km², a ratio of 1/4.

In view of the basin area, a ratio 1/3 is assumed. The duration of the flood is thus

$$t = t_c + t_k = t_c + 3t_c = 2.9 + 8.7 = 11.6 \text{ h}.$$

The volume of the pondage storage V can be obtained from the level--volumes curve of the reservoir.

Fig. 5.4a indicates the routing storage from elevation 103.00 to elevation 104.30, corresponding to a volume of 136 000 m³. The volume of the flood flow is calculated on the assumption of a triangular simplification of the hydrograph (*Fig. 5.5b*):

$$A = \frac{Qt}{2} = \frac{29.2 \times 11.6 \times 3600}{2} = 609\ 695 \text{ m}^3.$$

Substituting these values in eq. (5.5):

$$Q_{red} = 29.2 - \frac{2 \times 136\ 000}{11.6 \times 3600} = 29.2 - 6.5 = 22.7 \text{ m}^3/\text{s}.$$

The passage of the flood through the reservoir may thus reduce the peak flow by 6.5 m³/s, or roughly 30 per cent. It is evident that such computation is very approximate and may be used only on small streams with a lack of data on inflow.

5.4 Modern systems for hydrological data collection, transmission, and processing

(a) The constantly growing economic importance of water resources management and design and, in particular, the needs of flow-control systems have in recent years brought far-reaching developments in the collection, transmission, and processing of hydrological data. Although the classical methods described in chapter 3 of this book will retain their importance in the majority of countries for many years to come — particularly in developing countries — modern systems are increasingly being put into operational use. The systems concerning meteorological data are the most advanced at present; a global data collection, transmission, and processing system, known by the name of World Weather Watch (WWW), is being installed for operational use all over the globe through international cooperation within the World Meteorological Organization.

Hydrological systems obviously have a more limited scope — basin-wide, nation-wide, or regional (international basins) systems do and will

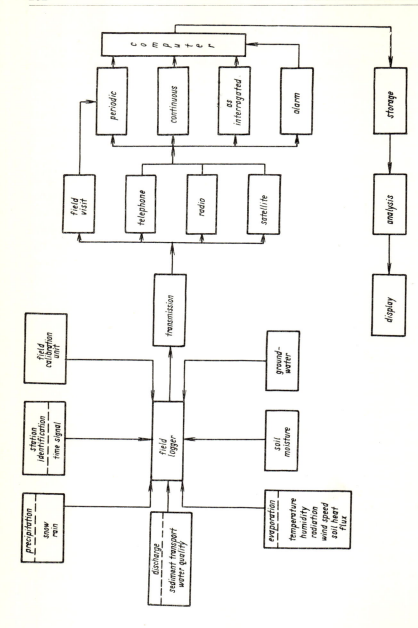

Fig. 5.7. Diagram of data system based on automatic hydrometeorological station

fulfill the needs of operational and engineering hydrology. However, before a system is implemented, these needs must be most carefully ascertained. In view of the considerable investment involved, advantages may only be obtained if the system design has been subjected to a cost-benefit evaluation.

In its simplest form, a modern system may be represented by a block diagram, as in *Fig. 5.7*. It is to be noted that the automatic hydrometeorological station on which this system is based integrates observations of almost all elements of the hydrological cycle. However, in practice such integration is not always feasible, owing, alas, not so much to technical or economic difficulties as to administrative and institutional problems. Such problems, encountered, for example, in countries where the meteorological, hydrological, and hydrogeological services are under separate administrations, may be conveniently overcome by joint planning and installation of the system, and eventually by a joint access to data storages installed in the different agencies' computerized data-processing centres. An example of this sort of joint planning is the effort of the British Natural Environment Research Council Working Party on Hydrometry and Instrumentation in co-operation with the Water Resources Board. An example of joint access to water resources data collected, transmitted, and processed by several systems is the Office of Water Data Co-ordination (OWDC) in the United States. Under this office's arrangements, the rainfall, evaporation, and other meteorological data, and data needed for flood forecasting, collected by the US Weather Bureau and stored in its data-processing centres, are directly accessible to other users, in particular the US Geological Survey which is in charge of collecting streamflow and groundwater data which are stored in its processing centre and are likewise directly accessible to others, in particular to the US Weather Bureau.

A joint hydrometeorological service, such as exists in Czechoslovakia, and the centralization of water resources activities, planning, and design under one authority greatly facilitates the establishment of basin-wide and national data systems. *Fig. 5.8* represents such a system for the basin of the River Vah in Czechoslovakia. The data system is designed for the operation and optimum multi-purpose management (power production, irrigation, flood control) of a river flow control system consisting, among other things, of 18 dams in a catchment of 17 630 km² with a total river length of 1925 km. An example of a nation-wide system with international links is given in *Fig. 5.9* which represents a block diagram of the data system used predominantly for flood forecasting by the Czechoslovak Hydrometeorological Institute (Hladny, 1966), with integrated meteorological and hydrological data collection, transmission, and processing.

From the above discussion, it appears that it is most difficult to give directives for the installation of modern data systems in hydrology, since these depend on the specific conditions of the catchment and the country.

However, the technical problems of the components of such a system, as represented in *Fig. 5.7*, are very similar in all systems and will be briefly described below.

(b) The *collection* (observational) component of the system presents two aspects: the network design and the instrumentation (sensor) equipment. Both aspects are interrelated and exercise a considerable feedback on their respective development.

The *network design*, as described in chapter 3, acquires a particular importance in a modern data system. The amount of continuously transmitted data becomes prohibitive as the network expands and a cost-benefit study is the most appropriate tool to solve this problem. The technical approaches, however, to network design, namely how many data-acquisition points are required, where to locate them, and for how long records should be collected, are still in the research stage. A recent international report by WMO (WMO/IHD Project Report No. 12, 1969) exposes this problem in great detail.

The *instrumentation* (observational) developments have recently made startling progress. It is to be noted that the sensor and local logger can often not be considered separately from the transmission (communication) component of the system. The following elements of both components must be considered.

Sensor: units (several physical element-sensing units sensing rainfall, air moisture, water stage, etc.), range, accuracy, timing of observation and transmission, display, power supply.

Converter: analogue (paper chart) or digital (paper tape or magnetic tape), signal coding, possibility of manual input.

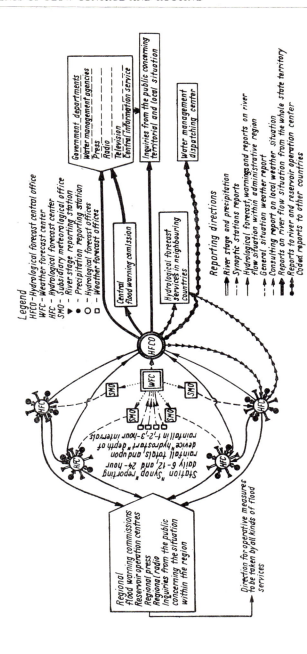

Fig. 5.9. Data system for nation-wide flood forecasting and warning (Czechoslovak Hydrometeorological Institute)

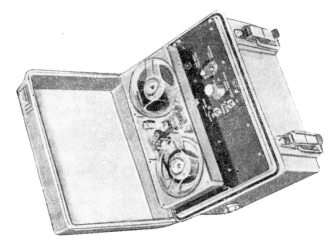

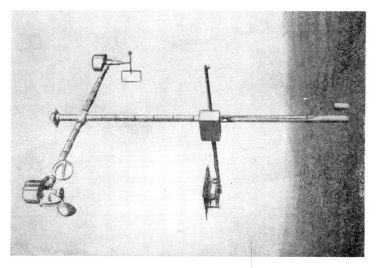

Fig. 5.10. Epsilon — UK Institute of Hydrology (Wallingford) climatological data-logging system

Transmission: pulse and/or frequency characteristics, transfer line, telephone line, phantom circuit, high-voltage line, radio (VHF), satellite retransmission and interrogation.

Received signal in centre: amplification, rectifying, conversion for processing (digital-digital), decoding.

Stations have been devised that measure a number of elements simultaneously; for example, water-level and water-quality parameters, rainfall, and the factors necessary for assessing evaporation. A station of this type has been developed in the UK, using a ten-channel battery-powered magnetic-type logger to record eight different parameters and time, at six-minute intervals. This station can operate for 14 days; then the tape cassette is replaced by a new one and taken to the computer at base for processing and analysis. The design philosophy behind this station was to produce a system cheaply enough for price not to prohibit its duplication in a river basin (*see Fig. 5.10*). In other words, several simple stations recording to a lower precision were preferred to one complex device, providing high precision at a single point.

Automatic measuring of river stage and discharges is also in the foreground of recent instrumentation development. In addition to the classical float and bubble-type manometer sensors, coupled to paper or magnetic-tape loggers, solid-state transducers have been developed (G. C. Dohler, 1969). They may be coupled to an advanced discharge-measurement system, such as ultrasonic pulses transmitted across the stream. On the basis of the Doppler principle, the travel time of the pulses is a function of the average stream flow velocity which may be computed from it.

A telemetering river stage station in full operational use is the 'limniphone', utilized by the Electricité de France hydrological service of France. The station in *Fig. 5.11* records on magnetic tape and, by a telephone line, automatically transmits river stages in hourly intervals or more frequently if required. An analogue display (on a stage-recorder paper chart) permits a suitable checking of the station's records. Rainfall observations may be included in this station also.

In many industrialized countries of the world, the *water-quality* monitoring constantly attracts increased attention. At present separate systems for water-pollution detection are installed, for later interconnection with the general water resources data systems. Such systems have been installed, for example, in the United States by the Ohio River Valley Water Sanitation Commission (ORSANCO).

Progress has been made in *telemetering* the information logged in the field. There are obvious advantages of VHF radio links over line telemetry, but the most flexible transmission system would be one operated through a satellite. In fact, one synchronous satellite might be sufficient to relay all the data that it could collect from a large part of the automatic hydrometeorological network of a medium-size country.

Fig. 5.11. Limniphone (automatic stage recording telemetering station).
Courtesy of Electricité de France, Grenoble, France

Satellites and aircraft are being employed already as vehicles for sensors of various types. Most use has been made of photography, and large numbers of photographs have already been employed for hydrological purposes. For example, various studies of the distribution of snow cover have been undertaken and the operational difficulties of mapping snow extent and depth have been investigated. These problems are described by Popham (1968), together with the other imaging techniques that could be utilized in snow surveillance and for other purposes. These include multicolour photometer mapping systems, multispectral photographic systems, image orthicon cameras, and panoramic scan systems, all designed to increase the breadth of information collected by the satellite. Other data-collecting systems rely on microwave scattero-meters, microwave radiometers and infra-red radiometers, while satellite-based radar systems are a distinct possibility. Studies of the correlation between sensor outputs and the hydrological characteristics of the terrain are under way and this subject is likely to prove one of the most difficult aspects of using these remote sensing techniques. For it is not too difficult to establish, in qualitative terms, the pattern of soil moisture or soil temperature variations over an area by an infra-red system, for example. It is another matter to confirm the 'ground truth' of these patterns. The same applies to ground-based radar measurement of precipitation, for, although radar has been employed for this purpose for some time, there remain some difficulties in interpreting and correlating radar measure-ments with those made in networks of standard gauges (Kessler and Wilk, 1968). Then there is the problem that, in any case, the standard gauge does not measure the true rainfall, and, what is more important, not even a good estimate of it on some occasions.

Paralleling these advances in instruments and data-collecting systems, there have been similar and possibly greater changes in the methods of data processing. Shelf after shelf of thick ledgers containing records of temperature, rainfall, or like factors are being replaced by magnetic discs or one of the other types of data store. In fact, the previously ubiquitous hand-written entry has given way to several forms of data presentation. Now, there is the choice between punch-cards, paper tape, magnetic tape, and various kinds of microfilm (WMO, 1969). Of course, each format requires its own specialized range of machines, including an electronic computer, and the staff to operate them.

(c) From the above, it appears that efforts to set up modern and automatic systems for hydrological data collection, transmission, and processing are under way in many countries of the world.

These efforts are internationally promoted and information on them is provided by the World Meteorological Organization, within its activities in hydrology, but mainly in connection with the Organization's system of World Weather Watch. As already mentioned, this is the first global data-monitoring, -communication, and -processing system which is operationally used. Its use and implications in hydrology and water resources management are described in WMO reports (Bruce and Nemec, 1967; Harding and Afanassiev, 1970) to which the reader is directed. Several other publications of WMO include detailed information on hydrological data systems, in particular the Technical Notes on 'Machine Processing of Hydrometeorological Data' (Isherwood et al., 1970) and on 'Automatic Equipment for Observing and Transmitting Hydrometeorological Elements' (Walser et al., 1971).

It is without doubt that the path followed by highly developed industrial countries in hydrological data systems will not entirely correspond to that followed by developing countries. However, it is already evident that many elements of these systems will immediately be made operational in the newly installed networks of these countries, without passing through the relatively cumbersome traditional practices.

(d) It may confidently be forecast that the problems of engineering hydrology may change basically with the advent of these new data systems. The availability, within the reach of each engineer, of software and hardware equipment processing all available data and their almost instantaneous display will no doubt considerably influence the hydrological activities of the engineer in the field and in the consulting office, to whom this book is addressed. Thus, this last section, describing modern data systems in hydrology, may conveniently serve as a forecast for future developments in engineering hydrology and as a conclusion to this book.

BIBLIOGRAPHY

Abbott, M. B.: *An Introduction to the Method of Characteristics,* Thames and Hudson, London, 1966.

Alekseyev, G. A.: *Grafoanaliticheskye sposoby opredeleniya i privedeniya k dlitelnomu periodu nablyudenij parametrov krivych raspredeleniya.* In: Trudy GGI, vyp. 73, pp. 90—140, 1960.

Alekseyev, G. A.: *Computation of Flood Flow* (in Russian), Gidrometeoizdat, Leningrad, 1955.

British Rainfall (yearbook), H. M. S. O., London.

British Rainfall 1939 (and subsequent years), H. M. S. O., London.

Bruce, J. P., Nemec, J.: *World Weather Watch and its Implications in Hydrology and Water Resources Management,* WMO/IHD Project Report No. 4, Geneva, 1967.

B. S. 3680, Part 3, 1964; Part 4, 1965.

Carter, R. W., Godfrey, R. G.: *Storage and Routing* in Manual of Hydrology *Part 3.* Flood Flow Techniques, USGS, USGPO, Washington 1960.

Chow, V. T: *Open Channel Hydraulics,* McGraw-Hill Book Company Inc., New York, 1959.

Chow, V. T. (editor): *Handbook of Applied Hydrology,* McGraw-Hill Book Company Inc., New York, 1964.

Crawford, N. H., Linsley, R. K.: *The Synthesis of Stream flow Hydrographs on a Digital Computer,* Tech. Rept. 12, Dep of Civ. Eng., Stanford University, 1962.

Criddle, W. D.: *Methods of Computing Consumptive Use of Water,* Paper 1507, Proc. Amer. Soc. Civ. Eng., 84, Jan. 1958.

Dohler, G. C.: *Automatic Hydrological Data Collecting Stations, in Instrumentation and Observing Techniques, Proceedings Hydrology Symposium No. 7,* NRC and Inland Waters Branch of Canada, Ottawa, 1969.

Dooge, J. C. I.: *A General Theory of the Unit Hydrograph,* Journ. Geophys. Res., vol. 64, No. 1, pp. 241—256, 1959.

Dub, O.: *Hydrology* (in Slovak), SNTL, Bratislava, 1963.

Dub, O., Nemec, J. (editors): *Manual of Hydrology* (in Czech), SNTL, Prague, 1969.

Dzubak, M., Nemec, J., Balco, M.: *Maximum Flood Estimation* (in Czech), Manual of Hydrology, Chapt. 7, SNTL, Prague, 1969

Estimated Soil Moisture Deficit over Gt. Britain: Explanatory Notes Meteorological Office, Bracknell (issued twice monthly).

Ezekiel, M.: *Methods of Correlation Analysis,* 2nd. ed., John Wiley and Sons Inc., New York, 1941.

Gould, B. W.: *Statistical Methods for Estimating the Design Capacity of Dams,* Journal of Engngs Australia 12/1961.

Guidebook on Nuclear Techniques in Hydrology, Technical Report No. 91 International Atomic Energy Agency, Vienna, 1968.

Hladny, J., Martinec, J., Duba, D.: *Hydrological Forecasts* in Hydrological Data for Water Resources Plannnig, Lecture Book, International post-graduate training course in hydrology, Prague, Czechoslovakia, 1966.

Hurst, H. E., Black, R. P., Simaika, Y. M.: *Long-Term Storage,* Constable, London, 1965.

Klemeš, V.: *Matrix Methods Applied to Runoff Control Computations* in Lecture Notes, volume 3, International Post-Graduate Training Course, 1968, Prague, Czechoslovakia.

Kostyakov, A. N.: *Elements of Irrigation and Drainage* (in Russian), GISL, Moscow, 1960.

Krickij, S. N., Menkel, M. F.: *Vodochozyaystvennyye raschoty.* Leningrad, 1952,

Langbein, W. B.: *Queing Theory and Water Storage,* Proceeding of the ASCE. Vol. 84, HY 5, 1958.

Linsley, R. K., Kohler, M. A., Paulhus, J. L. H.: *Applied Hydrology,* McGraw-Hill Book Company Inc., New York, 1949.

Linsley, R. K., Kohler, M. A., Paulhus, J. L. H.: *Hydrology for Engineers,* McGraw-Hill Book Company Inc., New York, 1959.

Logarithmic Plotting of Stage-discharge Observations, Tech. Note 3. Water Resources Board, Reading, 1966.

Lutcheva, A. A.: *Practical Hydrology* (in Russian), Gidrometeoizdat, Leningrad, 1950.

Manual of British Water Engineering Practice, Institution of Water Engineers, 4th ed., Vol. II, Chapts. 1, 2, London, 1969.

Moran, P. A. P.: *A Probability Theory of Dams and Storage Systems,* Austral. Journ. of Appl. Sci., Vol. 5, 1954.

Morgan, H. D.: *Estimation of Design Floods in Scotland and Wales, Paper No. 3 Symposium on River Flood Hydrology,* Instn. Civ. Engrs., London, 1966.

Nash, J. E.: *Applied Flood Hydrology,* in Proceedings Flood Hydrology Symposium Institution of Civil Engineers, London, 1968

Nash, J. E.: *Systematic Determination of Unit Hydrograph Parameters,* Journ. Geophys. Res., Vol. 64, No. 1 pp. 111—115, Jan. 1959.

Nash, J. E.: *A Unit Hydrograph Study, with Particular Reference to British Catchments,* Proc. Instn. Civ. Engrs., Vol. 17, p. 249, 1960.

Němec, J., Patočka, C.: *Elements of Hydrology* (in Czech), SZN, Prague, 1956.

Němec, J.: *Hydrology* (in Czech), SZN, Prague, 1965.

Němec, J.: *Simultaneous Use of an Analogue Single-Purpose Electronic Computers and a Physical Model of a Watershed,* Proceeding of Symposium on Computers, IASH Pub. No. 80, pp. 10—14, Gentbrugge, 1968.

Němec, J., Zezulak, J.: Digital, Analogue and Physical Simulation of Surface Runoff Process in a Catchment, Lecture Notes, International Courses in Hydraulic and Sanitary Engineering, Delft, Netherlands, 1971.

Nordenson, T. J.: *Preparation of Co-ordinated Precipitation, Runoff and Evaporation Maps,* Reports on WMO/IHD Projects, Report No. 6, Secretariat of the World Meteorological Organization, Geneva 1968.

Observer's Handbook, M. O. 554, Meteorological Office, HMSO, London, 1969.

Penman, H. L.: *Vegetation and Hydrology,* Technical Communication No. 53, Commonwealth Bureau of Soils, Harpenden, 1963.

Popham, R. W.: *Satellite Application to Snow Hydrology,* WMO/IHD Report No. 7, WMO Geneva, 1968.

Popov, E. G.: *Elements of Hydrological Forecasting* (in Russian), Gidrometeoizdat, Leningrad, 1968.

Potapov, M. V.: *Sochineniya,* 3. vol., Moskva 1st d. 1950, 398 p., 2nd ed. 1951, 480 p.

Rainbird, A. F.: *Methods of Estimating Areal Average Precipitation,* Reports on WMO/IHD Projects, Report No. 3, Secretariat of the World Meteorological Organization, Geneva, 1967.

Réménieras, G.: *Hydrologie de l'Ingénieur* (in French) Eyrolles, Paris, 1960.

Richards, B. D.: *Flood Estimation and Control,* 2nd ed., Chapman Hall, London, 1950.

River Flood Hydrology, Proceedings of the Symposium organized by the Institution of Civil Engineers, London, 1966.

Roche, M.: *Hydrologie de Surface* (in French), Gauthier—Villars, Paris, 1963.

Savarenskij, A. D.: *Metod raschota regulirovaniya stoka.* — "Gidrot. strojit." 1940, no. 2.

Standards for Methods and Records of Hydrologic Measurements, WMO/ECAFE publication, Flood Control Series No. 6, United Nations 1954.

Surface Water Yearbook of Great Britain, H. M. S. O., London.

Svanidze, G. G.: *Osnovy raschota regulirovaniya rechnogo stoka metodom Monte-Karlo,* Tbilisi, 1964.

Svoboda, A.: *Surface Runoff* (in Czech) in Manual of Hydrology, Chapt. 6, SNTL, Prague, 1969.

Symposium on Design of Hydrological Networks, Quebec, WMO/IASH publication.

Thorn, R. B. (editor): *River Engineering and Control Works,* Butterworth, London, 1966.

Todd, D. K.: *Ground Water Hydrology,* John Wiley & Sons Inc., New York, 1959.

Voskresenskyj, K. P.: *Hydrological Design of Projects on Small Rivers* (in Russian), Gidrometeoizdat, Leningrad, 1956.

Wilson, E. M.: *Engineering Hydrology,* Macmillan, London, 1969.

Wittmann, H.: *Hydrometeorology* (in German) in Manual for Civil Engineers, Chapt. 25, Springer Verlag, Berlin, Heidelberg, 1955.

294

WMO/IHD Project Report No. 12 (Rodda, J., W. B. Langbein., A. G. Kovzelj., D. R. Dawdy and K. Szesztay) *Hydrological Network Design, Needs, Problems, and Approaches,* WMO Secretariat, Geneva, 1969.

WMO, 1969, *Data Processing for Climatological Purposes,* Technical Note No. 100, WMO, Geneva.

World Meteorological Organization: *Guide to Climatological Practices,* WMO — No. 100. T. P. 44, Geneva, 1960.

World Meteorological Organization: *Guide to Hydrometeorological Practices,* 2nd ed., WMO — No. 168., Geneva, 1970.

World Meteorological Organization: *Guide to Meteorological Instrument and Observing Practices,* 2nd ed., WMO — No. 8. TP. 3, Geneva, 1961.

World Meteorological Organization: *Organization of Hydrometeorological and Hydrological Services,* Reports on WMO/IHD Projects, Report No. 10, Secretariat of the World Meteorological Organization, Geneva, 1969.

World Meteorological Organization: *Manual for Depth-Area-Duration Analysis of Storm Precipitation,* WMO — No. 237. TP. 129, Geneva, 1969.

World Meteorological Organization: *Measurement and Estimation of Evaporation and Evapotranspiration.* Technical Note No. 83, WMO — No. 201. TP. 105, Geneva, 1966.

World Meteorological Organization: *Hydrological Forecasting,* Technical Note No. 92, WMO — No. 228. TP. 122, Geneva, 1969.

World Meteorological Organization: *Practical Soil Moisture Problems in Agriculture,* Technical Note No. 97, WMO — No. 235. TP. 128, Geneva, 1968.

World Meteorological Organization: *Estimation of Maximum Floods,* Technical Note No. 98, WMO — No. 233. TP. 126, Geneva, 1969.

World Meteorological Organization: *Machine Processing of Hydrometeorological Data,* Technical Note No. 115, WMO No. 275, Geneva. 1971.

World Meteorological Organization: *Automatic Collection and Transmission of Hydrological Observations,* Technical Note (in print), WMO, Geneva, 1972.

World Meteorological Organization: *Technical Regulations,* 3rd ed., WMO — No. 49. Geneva, Volume 3 "Operational Hydrology".

Wundt, W.: *Hydrology* (in German), Springer Verlag, Berlin, Heidelberg, 1963.

ACKNOWLEDGEMENT

The author wishes to acknowledge with gratitude the permission granted by the World Meteorological Organization, through its Secretary-General, Mr D. A. Davies, to use material included in its valuable publications, the authors of which are well-known as hydrologists throughout the world.

Appendix 1(a). Basic units of measure of the MKSA system of units with some equivalents

Length

1 kilometre = 1000 metre   = 10^3 metre
1 centimetre = 0.01 metre   = 10^{-2} metre
1 millimetre = 0.001 metre   = 10^{-3} metre
1 micron   = 0.000001 metre   = 10^{-6} metre
1 angstrom   = 0.000000001 metre   = 10^{-10} metre
1 kilometre = 0.621 mile
1 mile   = 1.610 kilometre
1 metre   = 3.281 feet
1 foot   = 0.305 metre
1 centimetre = 0.3937 inch
1 inch   = 2.54 centimetre

Mass

1 kilogram = 1000 gram
1 kilogram = 2.202 pound
1 pound   = 0.4536 kilogram

Time

1 tropical (mean, solar, ordinary) year = 365.242198 day
1 day = 24 hour = 1440 minute = 86400 second

Force

1 newton   = 100.000 dyne = 10^5 dyne

Pressure

1 bar   = 10^5 newton/square metre   = 10^5 newton . metre^{-2}
1 bar   = 10^6 dyne/square centimetre   = 10^6 dyne . cm^{-2}
1 millibar = $\frac{1}{1000}$ bar = 10^2 newton . metre^{-2} = 10^3 dyne . cm^{-2}

Energy

1 joule   = 10,000,000 erg = 10^7 erg
1 15 °C calorie = 4.1855 joule

Energy per unit area

1 langley = 1 15 °C calorie . cm^{-2} = 4.1855 . 10^4 joule . metre^{-2}

Power

1 watt = 1 joule . sec^{-1}

Temperature

The Celsius temperature "t" of a system is given by
$$t = (T - 273.15)°C,$$
in which T is the temperature in degrees Kelvin.

Power

1 watt = 1 joule . sec^{-1}

Temperature

The Celsius temperature "t" of a system is given by
$$t = (T - 273.15)°C,$$
in which T is the temperature in degrees Kelvin.

Appendix 1(b). Recommended units for hydrological and meteorological elements. Commonly used alternative units and corresponding factors for conversion to recommended units are also shown*

Element (1)	Recommended unit (2)	Alternative units (3)	Factor for conversion from alternative unit (3) to recommended unit (2) (4)
Water-level (stage)	cm	ft	30.5
Stream discharge	m³/sec	cfs	0.0283
Unit discharge	l/sec km²	cfs/mile²	0.0103
Volume (storage)	m³	ft³	0.0283
		ac . ft	1230
		cfs . days	2450
Runoff depth	mm	in	25.4
Precipitation	mm	in	25.4
Precipitation intensity	mm/h	in/h	25.4
Snow depth	cm	in	2.54
Snow cover, area	%		
Water equivalent of snowpack	mm	in	25.4
Ice thickness	cm	in	2.54
Evaporation	mm	in	25.4
Evapotranspiration	mm	in	25.4
Soil moisture	%, volume	%, weight	(conversion depends on density)
Soil-moisture deficiency	mm	in	25.4
Sediment discharge	t/day	tons/day	0.907
Sediment concentration	kg/m³	ppm	(conversion depends on density)
Chemical quality	ppm		
Energy (heat)	cal (gramme)	Btu	252
Radiation	cal/cm²	ly	
Radiation intensity	cal/cm² min	ly/min	
Sunshine	% possible	h	(conversion depends on possible sunshine)

* Excerpt from the WMO Guide to Hydrometeorological Practices

Element	Recommended unit	Alternative units	Factor for conversion from alternative unit (3) to recommended unit (2)
(1)	(2)	(3)	(4)
Temperature	°C	°F	5/9 (°F − 32)
Wind speed	knots	mile/h	0.868
	m/sec		
Relative humidity	%		
Vapour pressure	mb	mm Hg	1.333
		in Hg	33.86
Atmospheric pressure	mb	mm Hg	1.333
		in Hg	33.86
Area	km²	mile²	2.59
		ac	0.00405
		ha	0.01

Note: Abbreviations used in the table are as follows:

ac — acre
Btu — British thermal unit
°C — degrees Celsius
cfs — cubic feet per second
cm — centimetre
°F — degrees Fahrenheit
ft — foot
ha — hectare

Hg — mercury
h — hour
in — inch
kg — kilogram
km — kilometre
l — litre
ly — langley
m — metre

mb — millibar
min — minute
mm — millimetre
ppm — parts per million by weight
sec — second
t — metric tonne

Appendix 1(c). Conversion tables for volume and discharge units used most frequently in hydrology

Conversion Table for Volume

Unit	Equivalents						
	Cu in.	Gal	Imperial gal	Cu ft	Cu m	Acre-ft	Sfd
Cubic inch	1	0.00433	0.00361	5.79×10^{-4}	1.64×10^{-5}	1.33×10^{-8}	6.70×10^{-9}
U.S. gallon	231	1	0.833	0.134	0.00379	3.07×10^{-6}	1.55×10^{-6}
Imperial gallon	277	1.20	1	0.161	0.00455	3.68×10^{-6}	1.86×10^{-6}
Cubic foot	1.728	7.48	6.23	1	0.0283	2.30×10^{-5}	1.16×10^{-5}
Cubic meter	61,000	264	220	35.3	1	8.11×10^{-4}	4.09×10^{-4}
Acre-foot	7.53×10^{7}	3.26×10^{5}	2.71×10^{5}	43.560	1230	1	0.504
Second-foot-day	1.49×10^{9}	6.46×10^{5}	5.38×10^{5}	86.400	2450	1.98	1

Conversion Table for Discharge

Unit	Equivalents						
	Gal/day	Cu ft/day	Gpm	Imperial gpm	Acre-ft/day	Cfs	Cu m/sec
U.S. gallon per day	1	0.134	6.94×10^{-4}	5.78×10^{-4}	3.07×10^{-6}	1.55×10^{-6}	4.38×10^{-5}
Cubic foot per day	7.48	1	5.19×10^{-3}	4.33×10^{-3}	2.30×10^{-5}	1.16×10^{-5}	3.28×10^{-7}
U.S. gallon per minute	1440	193	1	0.833	4.42×10^{-3}	2.23×10^{-3}	6.31×10^{-5}
Imperial gallon per minute	1728	231	1.20	1	5.31×10^{-3}	2.67×10^{-3}	7.57×10^{-5}
Acre-foot per day	3.26×10^{5}	43,560	226	188	1	0.504	0.0143
Cubic foot per second	6.46×10^{5}	86,400	449	374	1.98	1	0.0283
Cubic meter per second	2.28×10^{7}	3.05×10^{6}	15,800	13,200	70.0	35.3	1

m \ n	15	16	17	18	19	20	21	22	23
1	4.55	4.27	4.02	3.80	3.61	3.43	3.27	3.13	2.99
2	11.0	10.4	9.76	9.25	8.76	8.33	7.94	7.59	7.26
3	17.5	16.5	15.5	14.7	13.9	13.2	12.6	12.0	11.5
4	24.0	22.5	21.2	20.1	19.1	18.1	17.3	16.5	15.8
5	30.5	28.6	27.0	25.5	24.2	23.0	22.0	21.0	20.1
6	37.0	34.8	32.7	31.0	29.4	28.0	26.6	25.4	24.4
7	43.5	40.8	38.5	36.4	34.5	32.8	31.3	29.9	28.6
8	50.0	47.0	44.2	41.8	39.7	37.8	36.0	34.4	32.9
9	56.5	53.0	50.0	47.2	44.8	42.6	40.6	38.8	37.2
10	63.0	59.2	55.8	52.8	50.0	47.5	54.3	43.3	41.5
11	69.5	65.2	61.5	58.2	55.2	52.5	50.0	47.8	45.7
12	76.0	71.4	67.3	63.6	60.3	57.3	54.7	52.2	50.0
13	82.5	77.5	73.0	69.0	65.5	62.2	59.4	56.7	54.2
14	89.0	83.5	78.8	74.5	70.6	67.2	64.0	61.2	58.5
15	95.5	89.6	84.5	79.9	75.8	72.0	68.7	65.6	62.8
16		95.7	90.2	85.3	80.9	77.0	73.4	70.1	67.1
17			96.0	90.8	86.1	81.9	78.0	74.6	71.4
18				96.2	91.2	86.8	82.7	79.0	75.6
19					96.4	91.7	87.4	83.5	79.9
20						96.6	92.1	88.0	84.2
21							96.7	92.4	88.5
22								96.9	92.7
23									97.0
24									
25									
26									
27									
28									
29									
30									
31									
32									
33									
34									

$$p = \frac{m - 0.3}{n + 0.4} \text{ per cent}$$

24	25	26	27	28	29	30	31	32	33	34
2.87	2.76	2.65	2.56	2.46	2.38	2.30	2.23	2.16	2.10	2.04
6.97	6.70	6.44	6.20	5.98	5.78	5.58	5.41	5.25	5.09	4.94
11.1	10.6	10.2	9.85	9.50	9.18	8.88	8.60	8.33	8.08	7.85
15.2	14.6	14.0	13.5	13.0	12.6	12.2	11.8	11.4	11.1	10.8
19.3	18.5	17.8	17.2	16.5	16.0	15.5	15.0	14.5	14.1	13.7
23.4	22.4	21.6	20.8	20.1	19.4	18.8	18.2	17.6	17.1	16.6
27.5	26.4	25.4	24.4	23.6	22.8	22.1	21.3	20.7	20.0	19.5
31.6	30.3	29.2	28.1	27.1	26.2	25.3	24.5	23.8	23.0	22.4
35.7	34.3	33.0	31.8	30.6	29.6	28.6	27.7	26.9	26.1	25.3
39.8	38.2	36.7	35.4	34.1	33.0	31.9	30.9	30.0	29.1	28.2
43.9	42.1	40.5	39.1	37.7	36.4	35.2	34.1	33.0	32.0	31.1
48.0	46.1	44.3	42.7	41.2	39.8	38.5	37.3	36.1	35.0	34.0
52.0	50.0	48.1	46.4	44.7	43.2	41.8	40.4	39.2	38.0	36.9
56.1	53.9	51.9	50.0	48.2	46.6	45.1	43.6	42.3	41.0	39.8
60.2	57.9	55.7	53.6	51.8	50.0	48.4	46.8	45.4	44.0	42.7
64.3	61.8	59.5	57.3	55.3	53.4	51.6	50.0	48.5	47.0	45.6
68.4	65.7	63.2	60.9	58.8	56.8	54.9	53.2	51.5	50.0	48.5
72.5	69.7	67.0	64.6	62.3	60.2	58.2	56.4	54.6	53.0	51.5
76.6	73.6	70.8	68.2	65.9	63.6	61.5	59.6	57.7	56.0	54.4
80.7	77.6	74.6	71.9	69.4	67.0	64.8	62.7	60.8	59.0	57.3
84.8	81.5	78.4	75.6	72.9	70.4	68.1	65.9	63.9	62.0	60.2
88.9	85.4	82.2	79.2	76.4	73.8	71.4	69.1	67.0	65.0	63.1
93.0	89.4	86.0	82.8	79.9	77.2	74.7	72.3	70.0	68.0	66.0
97.1	93.3	89.8	86.5	83.5	80.6	77.9	75.5	73.1	70.9	68.9
	97.2	93.6	90.1	87.0	84.0	81.2	78.7	76.2	73.9	71.8
		97.4	93.8	90.5	87.4	84.5	81.9	79.3	77.0	74.7
			97.4	94.0	90.8	87.8	85.0	82.4	80.0	77.6
				97.5	94.2	91.1	88.2	85.5	82.9	80.5
					97.6	94.4	91.4	88.6	85.9	83.4
						97.7	94.6	91.7	88.9	86.3
							97.8	94.8	92.0	89.2
								97.8	95.0	92.2
									97.9	95.1
										98.0

Appendix 3. Coefficients S and S_1

C_s	$\dfrac{x_p - \bar{x}}{\sigma_x} = \Phi(p, C_s)$					$\Phi_5 - \Phi_{95}$
	Φ_5	Φ_{10}	Φ_{50}	Φ_{90}	Φ_{95}	
0.0	1.64	1.28	0.00	−1.28	−1.64	3.28
0.1	1.67	1.39	−0.02	−1.27	−1.61	3.28
0.2	1.70	1.30	−0.03	−1.26	−1.58	3.28
0.3	1.72	1.31	−0.05	−1.24	−1.55	3.27
0.4	1.75	1.32	−0.07	−1.23	−1.52	3.27
0.5	1.77	1.32	−0.08	−1.22	−1.49	3.26
0.6	1.80	1.33	−0.10	−1.20	−1.45	3.25
0.7	1.82	1.33	−0.12	−1.18	−1.42	3.24
0.8	1.84	1.34	−0.13	−1.17	−1.38	3.22
0.9	1.86	1.34	−0.15	−1.15	−1.35	3.21
1.0	1.88	1.34	−0.16	−1.13	−1.32	3.20
1.1	1.89	1.34	−0.18	−1.10	−1.28	3.17
1.2	1.92	1.34	−0.19	−1.08	−1.24	3.16
1.3	1.94	1.34	−0.21	−1.06	−1.20	3.14
1.4	1.95	1.34	−0.22	−1.04	−1.17	3.12
1.5	1.96	1.33	−0.24	−1.02	−1.13	3.09
1.6	1.97	1.33	−0.25	−0.99	−1.10	3.07
1.7	1.98	1.32	−0.27	−0.97	−1.06	3.04
1.8	1 99	1 32	−0 28	−0.94	−1.02	3.01
1.9	2.00	1.31	−0.29	−0.92	−0.98	2.98
2.0	2.00	1.30	−0.31	−0.90	−0.95	2.95
2.1	2.01	1.29	−0.32	−0.87	−0.91	2.92
2.2	2.02	1.27	−0.33	−0.84	−0.88	2.89
2.3	2.01	1 26	−0 34	−0.82	−0.85	2.86
2.4	2.00	1.25	−0.35	−0.79	−0.82	2.82
2.5	2.00	1.23	−0.36	−0.77	−0.79	2.79
2.6	2.00	1.21	−0.37	−0.75	−0.76	2.76
2.7	2.00	1.19	−0.38	−0.72	−0.74	2.74
2.8	2.00	1 18	−0 39	−0.70	−0 71	2.71
2.9	1.99	1 15	−0 39	−0.68	−0.69	2.68
3.0	1.97	1.13	−0.40	−0.66	−0.67	2.64
3.1	1.97	1.11	−0.40	−0.64	−0.64	2.62
3.2	1.96	1.09	−0.41	−0.62	−0.62	2.59
3.3	1.95	1 08	−0 41	−0.60	−0.60	2.56
3.4	1.94	1.06	−0.41	−0.59	−0.59	2.53
3.5	1.93	1.04	−0.41	−0.57	−0.57	2.50
3.6	1.93	1.03	−0.42	−0.56	−0.56	2.48
3.7	1.91	1.01	−0.42	−0.54	−0.54	2.45
3.8	1.90	1.00	−0.42	−0.53	−0.53	2.43
3.9	1.90	0.98	−0.41	−0.51	−0.51	2.41
4.0	1.90	0.96	−0.41	−0.50	−0.50	2.40
4.1	1.89	0.95	−0.41	−0.49	−0.49	2.38
4.2	1.88	0.93	−0.41	−0.48	−0.48	2.36
4.3	1.87	0.92	−0.40	−0.47	−0.47	2.34
4.4	1.86	0.91	−0.40	−0.46	−0.46	2.32
4.5	1.85	0.89	−0.40	−0.45	−0.45	2.30
4.6	1.84	0.87	−0.40	−0.44	−0.44	2.28
4.7	1.83	0.85	−0,40	−0.43	−0.43	2.26
4.8	1.81	0.82	−0.39	−0.42	−0.42	2.23
4.9	1.80	0.80	−0.39	−0.41	−0.41	2.21
5.0	1.78	0.78	−0.38	−0.40	−0.40	2.18
5.1	1.76	0.76	−0.38	−0.39	−0.39	2.15
5.2	1.74	0.73	−0.37	−0.38	−0.38	2.15

for C_S (Quantils method of Alexeyev)

$S = \dfrac{x_5 + x_{95} - 2x_{50}}{x_5 - x_{95}}$	$\Phi_5 + \Phi_{10} - \Phi_{90} - \Phi_{95}$	$S_1 = \dfrac{x_5 + x_{10} + x_{90} + x_{95} - 4x_{50}}{x_5 + x_{10} - x_{90} - x_{95}}$
0.00	5 84	0.00
0.03	5.84	0.03
0.06	5.84	0.05
0.08	5.82	0.08
0.11	5.82	0.10
0.14	5.80	0.12
0.17	5.78	0.15
0.20	5.75	0.18
0.22	5 73	0.20
0.25	5.70	0.23
0.28	5.67	0.25
0.31	5.61	0.28
0.34	5.58	0.31
0.37	5.54	0.34
0,39	5.50	0.36
0,42	5.44	0.39
0.48	5.39	0.41
0.45	5.33	0.44
0.51	5.27	0.47
0.54	5.21	0.49
0.57	5.15	0.52
0.59	5.08	0.55
0.63	5.02	0.58
0.64	4.94	0.60
0.67	4.86	0.62
0.69	4.79	0.65
0.72	4.72	0.67
0.74	4.65	0.70
0.76	4.59	0.72
0.78	4.51	0.74
0.80	4.43	0.76
0.81	4.37	0.78
0.83	4.30	0.80
0.85	4.24	0.82
0.86	4.17	0.73
0.87	4.11	0.84
0.89	4.07	0 86
0.90	4.00	0.88
0.91	3.95	0.89
0.92	3.91	0.90
0.92	3.86	0.91
0.93	3.81	0.92
0.94	3.76	0.93
0.94	3.72	0.93
0.95	3.68	0.94
0.96	3.63	0.95
0.97	3.58	0.96
0.97	3.53	0.97
0.98	3.46	0.97
0.98	3.42	0.97
0.98	3.36	0.98
0.98	3.30	0.98
0.98	3.24	0.98

Appendix 4. Co-ordinates

C_s	Probability										
	0.01	0.1	0.5	1	2	3	5	10	20	25	30
0.0	3.72	3.09	2.58	2.33	2.02	1.88	1.64	1.28	0.84	0.67	0.52
0.1	3.94	3.23	2.67	2.40	2.11	1.92	1.67	1.29	0.84	0.66	0.51
0.2	4.16	3.38	2.76	2.47	2.16	1.96	1.70	1.30	0.83	0.65	0.50
0.3	4.38	3.52	2.86	2.54	2.21	2.00	1.72	1.31	0.82	0.64	0.48
0.4	4.61	3.66	2.95	2.61	2.26	2.04	1.75	1.32	0.82	0.63	0.47
0.5	4.83	3.81	3.04	2.68	2.31	2.08	1.77	1.32	0.81	0.62	0.46
0.6	5.05	3.96	3.13	2.75	2.35	2.12	1.80	1.33	0.80	0.61	0.44
0.7	5.28	4.10	3.22	2.82	2.40	2.15	1.82	1.33	0.79	0.59	0.43
0.8	5.50	4.24	3.31	2.89	2.45	2.18	1.84	1.34	0.78	0.58	0.41
0.9	5.73	4.38	3.40	2.96	2.50	2.22	1.86	1.34	0.77	0.57	0.40
1.0	5.96	4.53	3.49	3.02	2.54	2.25	1.88	1.34	0.76	0.55	0.38
1.1	6.18	4.67	5.38	3.09	2.58	2.28	1.89	1.34	0.74	0.54	0.36
1.2	6.41	4.81	3.66	3.15	2.62	2.31	1.92	1.34	0.73	0.52	0.35
1.3	6.64	4.95	3.74	3.21	2.67	2.34	1.94	1.34	0.72	0.51	0.33
1.4	6.87	5.09	3.83	3.27	2.71	2.37	1.95	1.34	0.71	0.49	0.31
1.5	7.09	5.28	3.91	3.33	2.74	2.39	1.96	1.33	0.69	0.47	0.30
1.6	7.31	5.37	3.99	3.39	2.78	2.42	1.97	1.33	0.68	0.46	0.28
1.7	7.54	5.50	4.07	3.44	2.82	2.44	1.98	1.32	0.66	0.44	0.26
1.8	7.76	5.64	4.15	3.50	2.85	2.46	1.99	1.32	0.64	0.42	0.24
1.9	7.98	5.77	4.23	3.55	2.88	2.49	2.00	1.31	0.63	0.40	0.22
2.0	8.21	5.91	4.30	3.60	2.91	2.51	2.00	1.30	0.61	0.39	0.20
2.1	—	6.04	4.38	3.65	2.94	2.53	2.01	1.29	0.59	0.37	0.18
2.2	—	6.14	4.46	3.68	2.95	2.54	2.02	1.27	0.57	0.35	0.16
2.3	—	6.26	4.52	3.73	2.98	2.57	2.01	1.26	0.55	0.32	0.14
2.4	—	6.37	4.59	3.78	3.02	2.60	2.00	1.25	0.52	0.29	0.12
2.5	—	6.50	4.66	3.82	3.05	2.62	2.00	1.23	0.50	0.27	0.10
2.6	—	6.54	4.71	3.86	3.08	2.23	2.00	1.21	0.48	0.25	0.085
2.7	—	6.75	4.80	3.92	3.10	2.64	2.00	1.19	0.46	0.24	0.070
2.8	—	6.86	4.86	3.96	2.12	2.65	2.00	1.18	0.44	0.22	0.057
2.9	—	7.00	4.91	4.01	3.12	2.66	1.99	1.15	0.41	0.20	0.041
3.0	—	7.10	4.95	4.05	3.14	2.66	1.97	1.13	0.39	0.19	0.027
3.1	—	7.23	5.01	4.09	3.14	2.66	1.97	1.11	0.37	0.17	0.010
3.2	—	7.35	5.08	4.11	3.14	2.66	1.96	1.09	0.35	0.15	−0.006
3.3	—	7.44	5.14	4.15	3.14	2.66	1.95	1.08	0.33	0.13	−0.022
3.4	—	7.45	5.19	4.18	8.15	2.66	1.94	1.06	0.31	0.11	−0.036
3.5	—	7.64	5.25	4.31	3.16	2.66	1.93	1.04	0.29	0.085	−0.049
3.6	—	7.72	5.29	4.24	3.17	2.66	1.93	1.03	0.28	0.064	−0.072
3.7	—	7.86	5.35	4.26	3.18	2.66	1.91	1.01	0.26	0.048	−0.084
3.8	—	7.97	5.40	4.29	3.19	2.65	1.90	1.00	0.24	0.032	−0.095
3.9	—	8.08	5.45	4.32	3.20	2.65	1.90	0.98	0.23	0.020	−0.11
4.0	—	8.17	5.50	4.34	3.20	2.65	1.90	0.96	0.21	0.010	−0.12
4.1	—	8.29	5.55	4.36	3.22	2.65	1.89	0.95	0.20	0.000	−0.13
4.2	—	8.38	5.60	4.39	3.24	2.64	1.88	0.93	0.19	−0.010	0.13
4.3	—	8.49	5.65	4.40	3.24	2.64	1.87	0.92	0.17	−0.021	−0.14
4.4	—	8.60	5.69	4.42	3.25	2.63	1.86	0.91	0.15	−0.032	−0.15
4.5	—	8.69	5.74	4.44	3.26	2.62	1.85	0.89	0.14	−0.042	−0.16
4.6	—	8.79	5.79	4.46	8.27	2.62	1.84	0.87	0.13	−0.052	−0.17
4.7	—	8.89	5.84	4.49	3.28	2.61	1.83	0.85	0.11	−0.064	−0.18
4.8	—	8.96	5.89	4.50	3.29	2.60	1.81	0.82	0.10	−0.075	−0.19
4.9	—	9.04	5.90	4.51	3.30	2.60	1.80	0.80	0.084	−0.087	−0.19
5.0	—	9.12	5.94	4.54	3.32	2.60	1.78	0.78	0.068	−0.099	−0.20
5.1	—	9.20	5.98	4.57	3.32	2.60	1.76	0.76	0.051	−0.11	−0.21
5.2	—	9.27	6.02	4.59	3.33	2.60	1.74	0.73	0.035	−0.12	−0.21

$$\Phi_p = \frac{k_{p\%} - 1}{C_v} \text{ for Pearson's IIIrd Type distribution}$$

ordinates

40	50	60	70	75	80	90	95	97	99	99.9	100
0.25	0.00	−0.25	−0.52	−0.67	−0.84	−1.28	−1.64	−1.88	−2.33	−3.09	—
0.24	−0.02	−0.27	−0.53	−0.68	−0.85	−1.27	−1.61	−1.84	−2.26	−2.95	−20.0
0.22	−0.03	−0.28	−0.55	−0.69	−0.85	−1.26	−1.58	−1.79	−2.18	−2.81	−10.0
0.20	−0.05	−0.30	−0.56	−0.70	−0.85	−1.24	−1.55	−1.75	−2.10	−2.67	−4.67
0.19	−0.07	−0.31	−0.57	−0.71	−0.85	−1.23	−1.52	−1.70	−2.03	−2.54	−5.00
0.17	−0.08	−0.33	−0.58	−0.71	−0.85	−1.22	−1.49	−1.66	−1.96	−2.40	−4.00
0.16	−0.10	−0.34	−0.59	−0.72	−0.85	−1.20	−1.45	−1.41	−1.88	−2.27	−33.3
0.14	−0.12	−0.36	−0.60	−0.72	−0.85	−1.18	−1.42	−1.57	−1.81	−2.14	−2.86
0.12	−0.13	−0.37	−0.60	−0.73	−0.86	−1.17	−1.38	−1.52	−1.74	−2.02	−2.50
0.11	−0.15	−0.38	−0.61	−0.73	−0.85	−1.15	−1.35	−1.47	−1.66	−1.90	−2.22
0.09	−0.16	−0.39	−0.62	−0.73	−0.85	−1.13	−1.32	−1.42	−1.59	−1.79	−2.00
0.07	−0.18	−0.41	−0.62	−0.74	−0.85	−1.10	−1.28	−1.38	−1.52	−1.68	−1.82
0.05	−0.19	−0.42	−0.63	−0.74	−0.84	−1.08	−1.24	−1.33	−1.45	−1.58	−1.67
0.04	−0.21	−0.43	−0.63	0.74	−0.84	−1.06	−1.20	−1.28	−1.38	−1.48	−1.54
0.02	−0.22	−0.44	−0.64	−0.73	−0.83	−1.04	−1.17	−1.23	−1.32	−1.39	−1.43
0.00	−0.24	−0.45	−0.64	−0.73	−0.82	−1.02	−1.13	−1.19	−1.26	−1.31	−1.33
−0.02	−0.25	−0.46	−0.64	−0.73	−0.81	−0.99	−1.10	−1.14	−1.20	−1.24	−1.25
−0.03	−0.27	−0.47	−0.64	−0.72	−0.81	−0.97	−1.06	−1.10	−1.14	−1.17	−1.18
−0.05	−0.28	−0.48	−0.64	−0.72	−0.80	−0.94	−1.02	−1.06	−1.09	−1.11	−1.11
−0.07	−0.29	−0.48	−0.64	−0.72	−0.79	−0.92	−0.98	−1.01	−1.04	−1.05	−1.05
−0.08	−0.31	−0.49	−0.64	−0.71	−0.78	−0.90	−0.95	−0.97	−0.99	−1.00	−1.00
−0.10	−0.32	−0.50	−0.64	−0.70	−0.76	−0.866	−0.914	−0.930	−0.945	−0.952	−0.952
−0.12	−0.33	−0.50	−0.64	−0.69	−0.75	−0.842	−0.882	−0.895	−0.905	−0.910	−0.911
−0.13	−0.34	−0.50	−0.63	−0.68	−0.74	−0.815	−0.850	−0.860	−0.867	−0.870	−0.870
−0.14	−0.35	−0.51	−0.62	−0.67	−0.72	−0.792	−0.820	−0.826	−0.830	−0.833	−0.833
−0.16	−0 36	−0 51	−0.62	−0.66	−0.71	−0.768	−0.790	−0.795	−0.800	−0.800	−0.800
−0.17	−0.37	−0.51	−0.61	−0.66	−0.70	−0.746	−0.764	−0.766	−0.770	−0.770	−0.770
−0.18	−0.38	−0.51	−0.61	−0.65	−0.68	−0.724	−0.736	−0.739	−0.740	−0.740	−0.740
−0.20	−0.39	−0.51	−0.60	−0.64	−0.67	−0.703	−0.711	−0.714	−0.715	−0.715	−0.715
−0.21	−0.39	−0.51	−0.60	−0.63	−0.65	−0.681	−0.689	−0.690	−0.690	−0.690	−0.690
−0.22	−0.40	−0.51	−0.59	−0.62	−0.64	−0.661	−0.665	−0.666	−0.666	−0.667	−0.667
−0.23	−0.40	−0.51	−0.58	−0.60	−0.62	−0.641	−0.645	−0.646	−0.646	−0.646	−0.646
−0.25	−0.41	−0.51	−0.57	−0.59	−0.61	−0.621	−0.625	−0.625	−0.625	−0.625	−0.625
−0.26	−0.41	−0.50	−0.56	−0.59	−0.58	−0.605	−0.606	−0.606	−0.606	−0.607	−0.607
−0.27	−0.41	−0.50	−0.55	−0.57	−0.58	−0.586	−0.587	−0.588	−0.588	−0.588	−0.588
−0.28	−0.41	−0.50	−0.54	−0.55	−0.56	−0.570	−0.571	−0.571	−0.571	−0.572	−0.572
−0.28	−0.42	−0.49	−0.54	−0.54	−0.54	−0.555	−0.556	−0.556	−0.556	−0.556	−0.556
−0.29	−0.42	−0.48	−0.52	−0.53	−0.54	−0.541	−0.541	−0.541	−0.541	−0.541	−0.541
−0.30	−0.42	−0.48	−0.51	−0.52	−0.52	−0.526	−0.526	−0.526	−0.526	−0.527	−0.527
−0.30	−0.41	−0.47	−0.50	−0.51	−0.51	−0.513	−0.513	−0.513	−0.513	−0.513	−0.513
−0.31	−0.41	−0.46	−0.49	−0.49	−0.50	−0.500	−0.500	−0.500	−0.500	−0.500	−0.500
−0.31	−0.41	−0 46	−0.48	−0.484	−0.486	−0.487	−0.487	−0.487	−0.488	−0.488	−0.488
−0.33	−0.41	−0.45	−0.47	−0.473	−0.475	−0.476	−0.476	−0.476	−0.477	−0.477	−0.477
−0.32	−0.40	−0.44	−0.46	−0.462	−0.465	−0.465	−0.465	−0.465	−0.465	−0.465	−0.465
−0.32	−0.40	−0.44	−0.451	−0.454	−0.455	−0.455	−0.455	−0.455	−0.455	−0.455	−0.455
−0.32	−0.40	−0.43	−0.442	−0.444	−0.445	−0.445	−0.445	−0.445	−0.445	−0.445	−0.445
−0.32	−0.42	−0.42	−0.432	−0.434	−0.435	−0.435	−0.435	−0.435	−0.435	−0.435	−0.435
−0.32	−0.42	−0.40	−0.424	−0.424	−0.426	−0.426	−0.426	−0.426	−0.426	−0.426	−0.426
−0.32	−0.39	−0.41	−0.416	−0.416	−0.416	−0.416	−0.416	−0.416	−0.417	−0.417	−0.417
−0.33	−0.368	−0.401	−0.407	−0.407	−0.407	−0.408	−0.408	−0.408	−0.408	−0.408	−0.408
−0.33	−0.380	−0.395	−0.399	−0.400	−0.400	−0.400	−0.400	−0.400	−0.400	−0.400	−0.400
−0.33	−0.376	−0.388	−0.391	−0.392	−0.392	−0.392	−0.392	−0.392	−0.392	−0.393	−0.393
−0.33	−0.370	−0.382	−0.384	−0.385	−0.385	−0.385	−0.385	−0.385	−0.385	−0.385	−0.385

Appendix 5. Frequency coefficients

C_v	C_s	\multicolumn{10}{c}{Probability}
		0.01	0.1	0.5	1	2	3	5	10	20	25
0.05	0.10	1.197	1.162	1.134	1.120	1.106	1.096	1.084	1.064	1.042	1.033
0.10	0.20	1.416	1.338	1.276	1.247	1.216	1.196	1.170	1.130	1.083	1.065
0.15	0.30	1.667	1.528	1.429	1.381	1.332	1.300	1.258	1.197	1.123	1.096
0.20	0.40	1.922	1.732	1.590	1.522	1.452	1.408	1.350	1.264	1.164	1.126
0.25	0.50	2.208	1.952	1.760	1.670	1.578	1.520	1.442	1.330	1.202	1.155
0.30	0.60	2.515	2.188	1.939	1.825	1.708	1.636	1.540	1.399	1.240	1.183
0.35	0.70	2.848	2.433	2.127	1.985	1.840	1.752	1.637	1.469	1.276	1.203
0.40	0.80	3.200	2.696	2.324	2.156	1.980	1.872	1.736	1.536	1.312	1.232
0.45	0.90	3.578	2.971	2.530	2.332	2.125	1.999	1.837	1.603	1.346	1.256
0.50	1.00	3.978	3.266	2.744	2.511	2.270	2.126	1.938	1.670	1.378	1.277
0.55	1.10	4.399	3.568	2.969	2 700	2 419	2.254	2.040	1.737	1.407	1.297
0.60	1.20	4.846	3.886	3.196	2.890	2.572	2.386	2.146	1.804	1.438	1.312
0.65	1.30	5.316	4.218	3.431	3.086	2.736	2.521	2.248	1.871	1.468	1.332
0.70	1.40	5.809	4.563	3.681	3.289	2.897	2.659	2.358	1.938	1.497	1.343
0.75	1.50	6.318	4 922	3 932	3.498	3.055	2.792	2.462	1.998	1.518	1.352
0.80	1.60	6.848	5.296	4.192	3.712	3.224	2.936	2.568	2.064	1.544	1.368
0.85	1.70	7.409	5.675	4.460	3.924	3.397	3.074	2.674	2.122	1.561	1.374
0.90	1.80	7.984	6.076	4.735	4.150	3.565	3.214	2.782	2.188	1.576	1.378
0.95	1.90	8.581	6.482	5.018	4.372	3.736	3.366	2.890	2.244	1.598	1.380
1.00	2.00	9.210	6.908	5.298	4.605	3.910	3.507	2.996	2.303	1.610	1.386
1.05	2.10		7.329	5.599	4.828	4.080	3.654	3.108	2.352	1.618	1.388
1.10	2.20		7.750	5.900	5.050	4.250	3.800	3.220	2.400	1.625	1.390
1.15	2.30		8.200	6.200	5,290	4 435	3.960	3.310	2.450	1.628	1.370
1.20	2.40		8.650	6.500	5.530	4.620	4.120	3.400	2.500	1.630	1.350
1.25	2.50		9.125	6.815	5.775	4.810	4.270	3.500	2.535	1.626	1.340
1.30	2.60		9.600	7.130	6.020	5.000	4.420	3.600	2.570	1.621	1.330
1.35	2.70		10.10	7.465	6.285	5.185	4.565	3.700	2.605	1.616	1.320
1.40	2.80		10.60	7.800	6.550	5.370	4.710	3.800	2.640	1.610	1.310
1.45	2.90		11.12	8.110	6.815	5.535	4.845	3.880	2.670	1.600	1.295
1.50	3.00		11.65	8.420	7.080	5.700	4.980	3.960	2.700	1.590	1.280
1.55	3.10		12.20	8.770	7.330	5.860	5.115	4.045	2.725	1.575	1.260
1.60	3.20		12.75	9.120	7.580	6.020	5.250	4.130	2.750	1.560	1.240
1.65	3.30		13.28	9.460	7.840	6.185	5.385	4.215	2.775	1.545	1.210
1.70	3.40		13.80	9.800	8.000	6.350	5.520	4.300	2.800	1.530	1.180
1.75	3.50		14.35	10.17	8.360	6.525	5.650	4.385	2.825	1.512	1.148
1.80	3.60		14.9	10.54	8.620	6.700	5.780	4.470	2.850	1.495	1.115
1.85	3.70		15.52	10.90	8.885	6.875	5.905	4.540	2.875	1.478	1.088
1.90	3.80		16.15	11.25	9.150	7.050	6.030	4.620	2.900	1.460	1.060
1.95	3.90		16.75	11.62	9.415	7.230	6.165	4.705	2.915	1.441	1.040
2.00	4.00		17.35	12.0	9.680	7.410	6.300	4.790	2.930	1.422	1.020
2.05	4.10		17.98	12 38	9.940	7.605	6.425	4.870	2.940	1.404	1.000
2.10	4.20		18.60	12.75	10.20	7.800	6.550	4.950	2.950	1.385	0.979
2.15	4.30		19.25	13.12	10.465	7.975	6.665	5.025	2.975	1.362	0.954
2.20	4.40		19.90	13.50	10.73	8.150	6.780	5.100	3.000	1.340	0.930
2.25	4.50		20.55	13.90	11.00	8.335	6.900	5.165	3.000	1.318	0.905
2.30	4.60		21.20	14.30	11.28	8.520	7.020	5.230	3.000	1.295	0.880
2.35	4.70		21.85	14.70	11.54	8.710	7.135	5.290	2.99	1.268	0.850
2.40	4.80		22.50	15.10	11.80	8.900	7.250	5.350	2.980	1.240	0.820
2.45	4.90		23.15	15.46	12.08	9.095	7.375	5.400	2.965	1.205	0.786
2.50	5.00		23.80	15.85	12.36	9.290	7.500	5.450	2.950	1.170	0.752
2.55	5.10		24.45	16.25	12.64	9.470	7.625	5.485	2.925	1.130	0.716
2.60	5.20		25.10	16.65	12.92	9.650	7.750	5.520	2.900	1.090	0.680

for Pearson's IIIrd Type distribution for $C_s = 2 C_v$

ordinates

30	40	50	60	70	75	80	90	95	97	99	99.9
1.026	1.012	0.999	0.986	0.974	0.966	0.958	0.936	0.920	0.908	0.888	0.852
1.050	1.022	0.997	0.972	0.945	0.931	0.915	0.874	0.842	0.821	0.782	0.719
1.072	1.030	0.992	0.955	0.916	0.895	0.872	0.814	0.768	0.738	0.685	0.600
1.094	1.038	0.986	0.938	0.886	0.858	0.830	0.754	0.696	0.660	0.594	0.492
1.115	1.042	0.980	0.918	0.855	0.820	0.788	0.695	0.628	0.585	0.510	0.400
1.132	1.048	0.970	0.898	0.823	0.784	0.745	0.640	0.565	0.517	0.436	0.319
1.150	1.048	0.958	0.874	0.790	0.748	0.702	0.587	0.503	0.451	0.366	0.251
1.164	1.048	0.948	0.852	0.760	0.708	0.656	0.532	0.448	0.392	0.304	0.192
1.180	1.050	0.932	0.829	0.726	0.672	0.618	0.482	0.392	0.338	0.253	0.145
1.190	1.044	0.918	0.803	0.691	0.634	0.574	0.436	0.342	0.288	0.206	0.107
1.198	1.038	0.901	0.774	0.659	0.593	0.532	0.395	0.296	0.241	0.164	0.076
1.210	1.030	0.885	0.748	0.622	0.556	0.496	0.352	0.256	0.202	0.130	0.052
1.214	1.026	0.864	0.720	0.590	0.519	0.454	0.311	0.220	0.168	0.103	0.038
1.217	1.014	0.846	0.692	0.552	0.489	0.419	0.272	0.181	0.139	0.076	0.027
1.225	1.000	0.820	0.662	0.520	0.452	0.385	0.235	0.152	0.108	0.055	0.018
1.224	0.984	0.800	0.632	0.488	0.416	0.352	0.208	0.120	0.088	0.040	0.008
1.221	0.974	0.770	0.600	0.456	0.388	0.312	0.176	0.099	0.065	0.031	0.006
1.216	0.955	0.748	0.568	0.424	0.352	0.280	0.154	0 088	0 046	0.019	0.002
1.209	0.934	0.724	0.544	0.392	0.316	0.250	0.126	0.069	0.040	0.012	0.002
1.204	0.916	0.693	0.511	0.357	0.288	0.223	0.105	0.051	0.030	0.010	0.001
1.190	0.893	0.666	0.480	0.328	0.264	0.199	0.090	0.040	0.023	0.0074	0.000
1.175	0.870	0.640	0.450	0.300	0.241	0.175	0.074	0.030	0.016	0.0047	0.000
1.160	0.850	0.610	0.420	0.275	0.217	0.152	0.062	0.023	0.012	0.0031	0.000
1.145	0.830	0.580	0.390	0.250	0.193	0.130	0.049	0.016	0.008	0.0015	0.000
1.128	0.805	0.550	0.362	0.226	0.170	0.112	0.040	0.012	0.0059	0.001	0.000
1.110	0.780	0.520	0.334	0.203	0.146	0.094	0.030	0.0086	0.0038	0.0005	0.000
1.095	0.752	0.490	0.308	0.179	0.126	0.080	0.023	0.0063	0.0025	0.0002	0.000
1.080	0.725	0 460	0 283	0.155	0.106	0.065	0.016	0.0040	0.0012	0.000	0.000
1.060	0.698	0.432	0.258	0.138	0.092	0.056	0.012	0.0030	0.0006	0.000	0.000
1.040	0.670	0.405	0.234	0.120	0.077	0.046	0.009	0.0020	0.000	0.000	0.000
1.015	0.638	0.378	0.212	0.105	0.066	0.038	0.007	0.001	0.000	0.000	0.000
0 990	0.605	0.350	0.190	0.090	0.056	0.030	0.005	0.000	0.000	0.000	0.000
0.964	0.575	0.325	0.170	0.078	0.048	0.024	0.0035	0.000	0.000	0.000	0.000
0.938	0.545	0.300	0.150	0.067	0.039	0.019	0.002	0.000	0.000	0.000	0.000
0.904	0.518	0.275	0.134	0.058	0.033	0.016	0.0012	0.000	0.000	0.000	0.000
0.870	0.490	0.250	0.117	0.048	0.027	0.012	0.0005	0.000	0.000	0.000	0.000
0.845	0.462	0.231	0.104	0.040	0.022	0.0094	0.0002	0.000	0.000	0.000	0.000
0.820	0.435	0.212	0.090	0.033	0.017	0.0068	0.000	0.000	0.000	0.000	0.000
0.791	0.408	0.194	0.080	0.028	0.014	0.0053	0.000	0.000	0.000	0.000	0.000
0.762	0.380	0.175	0.070	0.022	0.011	0.0038	0.000	0.000	0.000	0.000	0.000
0.740	0.360	0.158	0.062	0.018	0.009	0.0025	0.000	0.000	0.000	0.000	0.000
0.719	0.340	0.140	0.053	0.014	0.007	0.0012	0.000	0.000	0.000	0.000	0.000
0.690	0.318	0.124	0.045	0.012	0.0054	0.0006	0.000	0.000	0.000	0.000	0.000
0.660	0.295	0.108	0.037	0.009	0.0038	0.000	0.000	0.000	0.000	0.000	0.000
0.635	0.278	0.095	0.032	0.0072	0.0029	0.000	0.000	0.000	0.000	0.000	0.000
0.610	0.260	0.082	0.027	0.0055	0.002	0.000	0.000	0.000	0.000	0.000	0.000
0.582	0.240	0.071	0.023	0.0043	0.0015	0.000	0.000	0.000	0.000	0.000	0.000
0.555	0.220	0.060	0.019	0.0031	0.002	0.000	0.000	0.000	0.000	0.000	0.000
0.528	0.200	0.055	0.016	0.0026	0.0006	0.000	0.000	0.000	0.000	0.000	0.000
0.500	0.180	0.050	0.012	0.0021	0.0002	0.000	0.000	0.000	0.000	0.000	0.000
0.475	0.165	0.040	0.010	0.0015	0.0001	0.000	0.000	0.000	0.000	0.000	0.000
0.450	0.150	0.040	0.008	0.001	0.000	0.000	0.000	0.000	0.000	0.000	0.000

Temperature [°C]	0.0	0.1	0.2	0.3	0.4	0.5	0.6	0.7	0.8	0.9
—9	2.32	2.30	2.28	2.27	2.25	2.23	2.21	2.20	2.18	2.16
—8	2.51	2.49	2.47	2.45	2.43	2.41	2.39	2.38	2.36	2.34
—7	2.71	2.69	2.67	2.65	2.63	2.61	2.59	2.57	2.55	2.53
—6	2.93	2.91	2.88	2.86	2.84	2.82	2.80	2.78	2.75	2.73
—5	3.16	3.13	3.11	3.09	3.06	3.04	3.02	2.99	2.97	2.95
—4	3.40	3.38	3.35	3.33	3.30	3.28	3.25	3.23	3.21	3.18
—3	3.67	3.64	3.62	3.59	3.56	3.53	3.51	3.48	3.46	3.43
—2	3.95	3.92	3.89	3.87	3.84	3.81	3.78	3.75	3.72	3.70
—1	4.26	4.22	4.19	4.16	4.13	3.10	4.07	4.04	4.01	3.98
—0	4.58	4.55	4.51	4.48	4.45	4.41	4.38	4.35	4.32	4.29
0	4.58	4.61	4.65	4.68	4.72	4.75	4.79	4.82	4.86	4.89
1	4.93	4.96	5.00	5.03	5.07	5.11	5.14	5.18	5.22	5.26
2	5.29	5.33	5.37	5.41	5.45	5.49	5.53	5.57	5.61	5.65
3	5.69	5.73	5.77	5.81	5.85	5.89	5.93	5.97	6.02	6.06
4	6.10	6.14	6.19	6.23	6.27	6.32	6.36	6.41	6.45	6.50
5	6.54	6.59	6.64	6.68	6.73	6.78	6.82	6.87	6.92	6.97
6	7.01	7.06	7.11	7.16	7.21	7.26	7.31	7.36	7.41	7.46
7	7.51	7.57	7.62	7.67	7.72	7.78	7.83	7.88	7.94	7.99
8	8.05	8.10	8.16	8.21	8.27	8.32	8.38	8.84	8.49	8.55
9	8.61	8.67	8.73	8.79	8.85	8.91	8.97	9.03	9.09	9.15
10	9.21	9.27	9.33	9.40	9.46	9.52	9.59	9.65	9.71	9.78
11	9.84	9.91	9.98	10.04	10.11	10.18	10.24	10.31	10.38	10.45
12	10.52	10.59	10.66	10.73	10.80	10.87	10.94	11.01	11.09	11.16
13	11.23	11.31	11.38	11.44	11.53	11.60	11.68	11.76	11.83	11.91
14	11.99	12.07	12.14	12.22	12.30	12.38	12.46	12.54	12.62	12.71

311

[Saturated vapour pressures in mm/Hg]

Tempe-rature [°C]	0.0	0.1	0.2	0.3	0.4	0.5	0.6	0.7	0.8	0.9
15	12.79	12.87	12.95	13.04	13.12	13.21	13.29	13.38	13.46	13.55
16	13.63	13.72	13.81	13.90	13.99	14.08	14.17	14.26	14.35	14.44
17	14.53	14.62	14.72	14.81	14.90	15.00	15.09	15.19	15.28	15.38
18	15.48	15.58	15.67	15.77	15.87	15.97	16.07	16.17	16.27	16.37
19	16.48	16.58	16.69	16.79	16.89	17.00	17.11	17.21	17.32	17.43
20	17.54	17.64	17.75	17.86	17.97	18.09	18.20	18.31	18.42	18.54
21	18.65	18.77	18.88	19.00	19.11	19.23	19.35	19.47	19.59	19.71
22	19.83	19.95	20.07	20.19	20.32	20.44	20.57	20.69	20.82	20.94
23	21.07	21.20	21.32	21.45	21.58	21.71	21.85	21.98	22.11	22.24
24	22.38	22.51	22.65	22.79	22.92	23.06	23.20	23.34	23.48	23.62
25	23.76	23.90	24.04	24.18	24.33	24.47	24.62	24.76	24.91	25.06
26	25.21	25.36	25.51	25.66	25.81	25.96	26.12	26.27	26.43	26.58
27	26.74	26.90	27.06	27.21	27.37	27.54	27.70	27.86	28.02	28.19
28	28.35	28.51	28.68	28.85	29.02	29.18	29.35	29.53	29.70	29.87
29	30.04	30.22	30.39	30.57	30.75	30.92	31.10	31.28	31.46	31.64
30	31.82	32.01	32.19	32.38	32.56	32.75	32.93	33.12	33.31	33.50
31	33.70	33.89	34.08	34.28	34.47	34.67	34.86	35.06	35.26	35.46
32	35.66	35.87	36.07	36.27	36.48	36.68	36.89	37.10	37.31	37.52
33	37.73	37.94	38.16	38.37	38.58	38.80	39.02	39.24	39.46	39.68
34	39.90	40.12	40.34	40.57	40.80	41.02	41.25	41.48	41.71	41.94
35	42.18	41.41	42.64	42.88	43.12	43.36	43.60	43.84	44.08	44.32
36	44.56	44.81	45.05	45.30	45.55	45.80	46.05	46.30	46.56	46.81
37	47.07	47.32	47.58	47.84	48.10	48.36	48.63	48.89	49.16	49.42
38	49.69	49.96	50.23	50.50	50.77	51.05	51.32	51.60	51.88	52.16
39	52.44	52.73	53.01	53.29	53.58	53.87	54.16	54.45	54.74	55.03

INDEX